The Lessening Stream

The Lessening Stream

An Environmental History of the Santa Cruz River

Michael F. Logan

The University of Arizona Press   Tucson

The University of Arizona Press
© 2002 The Arizona Board of Regents

LIBRARY OF CONGRESS CATALOGING-IN-PUBLICATION DATA
Logan, Michael F., 1950–
The lessening stream : an environmental history of the Santa Cruz River /
Michael F. Logan.
p. cm.
Includes bibliographical references (p.) and index.
ISBN-13: 978-0-8165-1586-8 (cloth : alk. paper)—ISBN-10: 0-8165-1586-7
ISBN-13: 978-0-8165-2605-5 (pbk. : alk. paper)—ISBN-10: 0-8165-2605-2
1. Human ecology—Santa Cruz River Valley (Ariz. and Mexico)—History.
2. Santa Cruz River (Ariz. and Mexico)—History. I. Title.
GF504.A6 L64 2002
333.91´6213´09791—dc21
2001004441

Publication of this book is made possible in part by a publication grant from the
Charles Redd Center for Western Studies at Brigham Young University.

11 10 09 08 07 06 6 5 4 3 2

to Lloyd R. Logan, Jr.
1921–1998

Contents

Illustrations

Acknowledgments

THIS BOOK STARTED WITH TWO fairly simple questions: when and why did the Santa Cruz River dry up? I knew the river had flowed through Tucson in the not-so-distant past and assumed that the answers to these questions would be straightforward and linear. The river flowed with an optimum volume at a certain point in time, but from that moment of rushing surplus, the flow decreased until the surface stream sank beneath the sand. The river's story from that point became a subsurface history, the water table declining, once again, in a simple linear fashion. However, as the following pages attest, the answers to my basic questions proved to be much more complex than I had imagined. What I thought at the outset would take a dozen pages to describe has taken more than two hundred pages to narrate.

The complexity of the answers—as it turns out, there is no single date or reason for the river's disappearance—has much to do with the invisibility of evidence and information pertaining to the questions. Most of the river's history took place before human beings lived in the valley, and most of the human presence in the valley preceded any written or recorded observations. More recently, the subsurface history of the river's aquifer has been obscured in a different fashion. Hydrologists can apply technological methods to characterize the movement and nature of underground water. Although the nature of the aquifer is made visible by their work, the hydrologists' pronouncements about the groundwater are often obscured by conceptual complexity and technical terminology. Hydrologists, anthropologists, and scholars from other disciplines have added to my understanding of the river's history, and in all cases their insights add complexity to the answers to my simple questions.

I owe a great debt of gratitude to those whose insights and analyses have broadened my understanding of the river, and to those whose help has contributed to the book in many ways. Hal Rothman encouraged me early and often. He helped me shape portions of two early chapters of the book into an article for *Environmental History*. In a more general sense, I always look forward to his latest report on the continuing saga of the Las Vegas boom. Many anonymous readers—of the book manuscript, related grant proposals, and article manuscript—provided helpful criticism and cogent suggestions. I cannot thank these readers specifically, but trust that those I fail to mention will understand my thanks extends to them as well.

The History Department at Oklahoma State University has been generous in extending research support to me, particularly in the form of summer travel support. Within the History Department, J. Paul Bischoff, friend and colleague, gave advice and guidance in many matters theoretical and technological. His knowledge of computers and graphics would allow him to make a fortune in private industry, but his commitment to the historian's profession keeps him toiling within the penurious realm of academia. The College of Arts and Sciences at OSU has also given me research support. In particular, Toni Shakley and Shelly Schultz helped me with grant applications and provided encouragement for this project from within the college's research funds. Beth McTernan at the Environmental Institute at OSU helped as well, providing guidance and support for this project. I want to especially thank Tanya Beer at the institute. Tanya inquired about my research and provided a spark of interest in the book when my own focus was drifting. She went on to write a clear and insightful summary of my research for the institute's newsletter. I also want to thank the Oklahoma Humanities Council for their generous support.

In Tucson many friends and scholars have assisted me in this project. The archives at the Arizona Historical Society are a treasure that many scholars appreciate and value. The state legislature, on the other hand, seems to think the archive can run on goodwill alone. Riva Dean and her staff at the archive provided help and support in countless ways. At the Arizona State Museum, Diana Hadley gave me the benefit of her kindness, insights, and support, and introduced me to others who provided key information for this study. In particular, Bill Doelle explained to me the latest archaeological assumptions about the Hohokam. Of course, any errors in my rendering of the prehistoric use of the river are my own. Friend and photographer Kent Kochenderfer graciously shared his thoughts and insights concerning the river's changing circumstances. Photographs by Kent appearing in this book were produced in collaboration with Richard McBain at Centric Photo in Tucson. Richard shares Kent's interest in the river's history and their photos and presentations attest to their keen appreciation for the desert landscape of southern Arizona.

Many hydrologists assisted me with this project. In Nogales, Ken Horton explained the history and current status of the Nogales, Arizona, water system. At the Tucson Water Company, Mitch Basefsky, Joe Marra, and Joe Herstel gave me information on the river in the Tucson Basin, and pointed me in helpful directions for information on the river's surface water and groundwater circumstances throughout the valley. Anna Cota-Robles helped me download images of Pete the Beak, the company's conservation mascot.

Conversations with Ron McEachern, Jack Long, and Doug Mason brought me up-to-date on the status of agriculture in the lower basin. I also benefitted from their explanations of the harsh economic realities of CAP water and of the evolving relationship between agriculture and urban, industrial, and tourist development. Clearly, any characterization of agriculture's use of water in the valley is incomplete without the insights provided by the farmers themselves. In part through these conversations about agriculture in the valley, I have come to the conclusion that no one in the valley is possessed of a purely white hat, nor are there any monolithic villains in the use of the valley's water resources.

Of course, any errors that remain in the book are my own. The help given me by those I have mentioned, along with many other friends and scholars, have made this book what it is: not the last word on the river's history, but an attempt to make understandable how the river got to this point in time.

At the University of Arizona Press this project received early and generous support from Joanne O'Hare. More recently, Christine Szuter and Patti Hartmann provided encouragement and assistance. Kirby Mittelmeier deserves praise for his careful editing of the manuscript. I also wish to acknowledge the publication grant awarded to the press by the Charles Redd Center for Western Studies.

Last, I want to thank my wife Patti and sons Owen and Bryce for their support and understanding while I worked on this project. They are the center of my life, much as the river is the center of life in the valley.

The Lessening Stream

Introduction
The Lessening Stream

IN 1910, G.E.P. SMITH, the eminent University of Arizona hydrologist, described the Santa Cruz River in the Tucson Basin as "ever a dwindling stream." He was studying the water sources in the basin and had come to the conclusion that the river had diminished to such an extent—a "brook" he called it—that its tributary, Rillito Creek, was by far the more promising source for water in the basin. The Santa Cruz River had provided water for residents along its banks for thousands of years, but in the early twentieth century places formerly manifesting surface water year-round ceased to flow, and the growing city of Tucson was forced to look beyond the river, and even beyond the basin, to secure sufficient water for its future.[1]

About thirty years prior to Smith's observation, my great-grandparents homesteaded 320 acres along the Santa Cruz River, twelve miles north of the international border with Mexico. She was a schoolteacher and he was a pharmacist, both from northern California. They came to Arizona in the 1880s to start fresh, acquire land of their own, and continue a family pioneering tradition. When deciding where to settle, the Santa Cruz River valley attracted them with the prospect of good grazing and fertile land. The allure was great in the moist 1880s, the grass high, and the water plentiful; a promise of success seductive enough to overcome even the threat of Apache raids. But success proved to be elusive—wet cycles always give way to dry ones. Nonetheless, at the heart of the enticing scene for my great-grandparents, and for countless other residents of the valley, was the river.

I mention this personal connection to the river for two reasons. First, to indicate to readers that my interest in the Santa Cruz River stems at least in part from familial traditions and personal experiences. My grandfather, mother, and I were born either within sight of the river, or within a few miles of the river's banks. Through my own observations, as well as through

family stories, I have long been conscious of the river's presence and changing circumstances.

The second reason I mention my great-grandparents is to illustrate a
primary goal of this book: to tell the story of the entire river, beyond a
narrow focus on the Tucson Basin, or any other particular stretch of the
river. Admittedly the Tucson Basin will come to dominate parts of the following narrative because it is home to the largest population in the river
valley and because it is where the river undergoes its most profound
changes during the modern era. Still, this environmental history seeks to
link developments throughout the river valley so that a more complete
picture of the river's history emerges. My interest and training in environmental history has lead me to conclude that the history of this river, so
significant for me personally, carries a much broader importance. The
history of this little river in southern Arizona explicates regional, national,
and international issues in environmental history.

Readers will find this book fits into the western history tradition of
Walter Prescott Webb. In *The Great Plains*, Webb offered an analytical
approach that started with the region's semiarid environment.[2] There is no
need here to delve into the historiographical debate over the most appropriate definition of the western region, or of western history. Taken as a
fundamental assumption of this book is the semiarid nature of the river
valley. The organizing framework for the book is "place," although processes of settlement and "frontier" cultures also play a role in the narrative.
As an intellectual starting point for this emphasis on place, the concept of
bioregionalism offers a valid methodological framework, especially for the
early chapters of this book. Bioregionalism[3] fosters a consciousness of watersheds as a defining parameter for ecological studies. This describes the
environment of the Santa Cruz River and its watershed in the surrounding
mountains (a description of the watershed and its physical parameters follows in chapter 1).[4]

For the later chapters of the book another model proved helpful. Theories concerning "hydraulic societies," such as those posed by Donald Worster in *Rivers of Empire*, provide a context for understanding the valley's
twentieth-century search for water resources. Specifically, the search for
new supplies of water in Tucson went beyond the river valley, laying claim
to the last of the large-scale reclamation projects in the West. The Central
Arizona Project (CAP) delivered Colorado River water to Tucson in 1990.
Arizona boosters, politicians, and engineers had conceived of the project
explicitly to deliver water across watersheds and bioregions. The era of

massive reclamation projects that dominated so much of western history in the twentieth century came to a close with the completion of the CAP and its terminus outside of Tucson.[5]

From its headwaters in Arizona's Canelo Hills, the Santa Cruz River travels through Mexico for thirty-two miles, the only river to originate within U.S. boundaries, flow out of the country, and then return to continue its course within the United States. This U-turn in the river's course signified little for most of the river's history, but the twentieth-century issues of economic development on the border, and pollution concerns that cross the border, cast the little river with the unique about-face in a new light. National policies on both sides of the border have affected the river, and likewise, the river itself has influenced international policy making and diplomacy.

As the reader will quickly notice, this work in environmental history relies on an interdisciplinary collection of sources. In no case do I claim expertise in all of the disciplines, from anthropology to zoology, that pertain to the history of the river, its aquifer, and its ecology. This book relies on scholars, scientists, and engineers from a wide variety of fields for their insights into the river's history. The notes are intended to allow readers to pursue their own interests by consulting the multi-disciplinary array of books, articles and essays, as well as archival material, that have served as source material for this book. I agree that the heart of environmental history lies in the humanities, but for academic accuracy and logical consistency, the discipline must keep both feet firmly grounded in the physical and social sciences. I do not presume that the physical sciences hold a monopoly on truthfulness and accuracy. Rather, one of the contributions of environmental history is to identify the subjective assumptions and at times narrow limits of vision that accompany scientific or engineering reports, and place those studies within a broader context of historical understanding.

Although environmental history places nature at the center of the historical narrative, human society remains an integral component of the story. Human decisions and lifestyles have impacted the Santa Cruz River and surrounding environment, and then bounced back to affect the human cultures in the river valley. The societies supported by the surface flow of the river varied greatly in their particulars, but existed within boundaries determined by the ecosystem.

For perhaps a half-million years prior to my relatives' arrival in the valley in the 1880s, and Smith's observation in 1910, the river had meandered, oozed, and crept 205 miles from its headwaters in the Canelo Hills to its terminus in the Gila River valley near Phoenix. The river drains a watershed of 8,581 square miles, an area slightly larger than Massachusetts.[6] In the hierarchy of rivers, particularly rivers in the predominantly arid Southwest, the Santa Cruz does not amount to much. It is a poor second cousin to the Colorado, a river that seems naturally to attract grand adjectives. No one ever described the Santa Cruz River as grand, although residents along its banks have always recognized its value and significance.

For the purposes of an environmental study, however, the Santa Cruz River's smallness proves a virtue. There is at least the chance of coming to understand the countless forces, processes, and variables affecting the diminutive river and its watershed. Much of the traditional scope of historical study fades from view when the focus shifts to causative agents of environmental rather than cultural or societal change. The river's size or economic weight becomes inconsequential. Of consequence is what happened to the river. What processes are noticeable, what forces are measurable? For example, the effect on the Santa Cruz River of several thousand cattle grazing in the hills and trampling the banks along its course is much greater than the imprint of a few hundred pre-industrial human beings living on its banks, be they Native American, Spanish, or Anglo-American. Thus, at least in the early chapters, the environmental history of the Santa Cruz River will pay little attention to governmental policies that fostered or inhibited colonization, the planting of missions, or the prosecution of the Apache wars, unless, for instance, those policies caused an increase in the size of the cattle herds roaming the valley. The Apache wars were of tremendous importance to the individuals and cultures involved, but may in fact pale to relative insignificance as far as the river is concerned.

Eventually, a culture appeared on the banks of the river that surpassed cattle and rivaled even climate in its influence on the river: the industrial society that arrived in the Tucson Basin with the Southern Pacific Railroad in 1880. With the steam locomotive came the prospect and soon the reality of the widespread use of steam-powered pumps to lift groundwater from the Santa Cruz River's shallow aquifer. The railroad also brought the prospect of wider markets for the region's products. Cattle and copper became linked to national and international markets. The river continued to be the center of life in the valley, but it now coursed through a much larger world. This second formulation of the river—underground and vast—is recent and current, and diverges widely in its influences on human cultures from the earlier, aboveground, slowly meandering river.

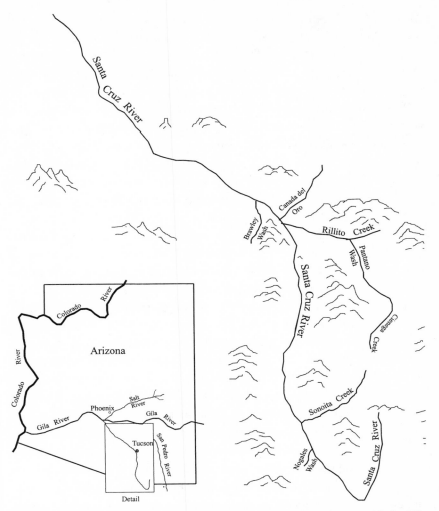

The Santa Cruz River flows 205 miles from its headwaters to its confluence with the Gila River near Phoenix, a short distance compared to the Colorado River's 1,440-mile journey from mountains to sea. These maps give some indication of the relative size of these rivers, although the lines on the maps are to some degree theoretical, since the Colorado reaches the sea only occasionally, and the Santa Cruz rarely flows over its entire course.

Previously the river had been used to the extent of its flow, but with the arrival of industrial technology, the river could be tapped beyond and below its flow. The transition is easily marked by a drop in the aquifer. A steady current and shallow groundwater support willow and cottonwood, with cattail *cienegas* (marshes) occurring where the water slows to a crawl. As water tables drop, cattails and willows disappear first. Mesquite bosques usually replace cottonwood ecosystems, since the mesquite taproots can readily seek out groundwater to depths of eighty feet or more. When the water table drops below even mesquite's ability to tap, the sage brush and creosote environment takes over. All of these transitions have taken place, and continue to take place, along the Santa Cruz River.

How inhabitants interpret these transitions depends on the perception of the people observing them. In my study of the river, I have concluded that three general types of perception have defined the residents of the river valley—archaic, modern, and postmodern—and I suggest a loosely chronological variance in these interpretations of the river's nature. Only in the most contemporary period does a consistent set of perceptions about the river fail to materialize.

By archaic, I suggest a view of the river that goes no farther than the water in the surface flow. This defines the human perception of the river through most of its history, from the first human presence in the valley to roughly the late nineteenth century. Native American, Spanish, Mexican, and Anglo-American inhabitants all viewed the river as providing sustenance within the narrow boundaries of its meager flow. Cultures tried to stretch or harness the river's flow with as much skill and technical expertise as they could bring to bear, but ultimately the river dictated the limits of its use. Given this power over human societies, it is not surprising that most of those bringing an archaic perception to the river also came to imbue the river's water with spiritual or mystical qualities. At the very least, this perception ascribes the water with characteristics approaching the anthropocentric: the river as pernicious, nurturing, treacherous, or generous.

By the modern perception, I mean the view of the river that looks beyond, through, and beneath the surface flow of the river to the aquifer underground. An understanding of the relationship between the surface flow and the percolation of groundwater defines this perception, and with it comes an appreciation for the relative abundance of the groundwater supply versus the scarce aboveground flow. A culture that digs wells may or may not indicate a modern vision of the river, since shallow wells might be dug in the river channel during times of drought with the understanding that the river's water has simply receded temporarily beneath the surface. Rather, the modern perception depends on a more complete view of the hydrological system that creates the surface flow of the river, and as such is generally accompanied by a faith and reliance on technological expertise in the management and exploitation of the water resource. Not without spirituality, this perception embodies the culture of the early 1900s, when science came to challenge religion in the hearts of many Americans. The modern temper held that the salvation of human society would come from science and technology, not from religion.

By postmodern I mean the more contemporary perception of the river, a generation or so old, a vision of the river's nature that has been attenuated to the point of extreme myopia. To most inhabitants of the river valley today, water is something that comes from a faucet, is delivered in bottles to the doorstep, or is purchased in six-packs at the grocery store. The river is a dry wash that sometimes floods during the monsoon or tropical storms, but the muddy, swift water that courses down the channel on those spectacular occasions is certainly not the same water that comes from the kitchen tap. This perception has much to do with assumptions that go far beyond the realm of intellectual awareness. Presumably, if you asked most Tucsonans to identify where their water came from, they would respond with some understanding of the river's aquifer or even the current status of CAP delivery. These expressions of understanding would be tenuous, however, vacillating between confused poles of perception: water in the river valley can be at once clear or muddy, healthful or tainted, cheap or prohibitively expensive.

Postmodernism's insight is perhaps best explained in the awareness it fosters that the river exists with no more certainty than its linguistic construction allows. Cultures use language to define their understandings of such things as rivers, and as cultures and languages change, so do meanings. The postmodern label is appropriate for the current society in the valley because the society is culturally diverse and politically variegated. Several opposing definitions and interpretations of the river's meaning and significance exist simultaneously in the valley, and no clear consensus about the river's fundamental nature exists. The chaotic discourse swirling around the river is characteristic of the contemporary postmodern era in the river's history.

During the period of the river's aboveground flow, consensus about its nature and significance was clearly present. Late in the nineteenth century the Santa Cruz River still flowed perennially, even if somewhat skimpily, along many reaches of its course. Now it is dry most of the time through most of its course and little remains of the cultures that inhabited the valley during the centuries and millennium of the river's aboveground existence. Hohokam pots grace museum shelves, with knock-offs available for tourist dollars. Spanish vestiges survive perhaps more genuinely in the Southwest, but with little of the true ambience of mission or presidio life. The restored interior of the mission at San Xavier del Bac inspires such awe, in part, because the Baroque artwork is so remote from early twenty-first-century concerns.

The flow and ebb in the recent past is but the latest example of the river's constant material circumstance. Ebb and flow describes the history of the river. No straight-line decline appears in the narrative, although many observers of the river have presumed such to be the case. Those observers assume a point in the past when the river flowed to its fullest extent, and from that point steady and inexorable atrophy describes the river's sad history. But the river's environmental history does not constitute a moral tale of social decline from a position of ecological grace to environmental depravity—a declension story. In fact no single moment of abundance exists in the river's history; rather, many such moments appear. The river dried up and came back several times, even within the last two hundred years. To phrase this idea archaically, the river has repeatedly flowed generously and shrunk to a meager tease. Likewise the aquifer's history fails to live up to a simple declension story. Just as the surface flow surged and shrank in the pre-industrial period, the water table has dropped and risen in the more recent period as natural recharge from rain and floods, along with varying rates of groundwater pumping, have caused the aquifer to variously sink or rise.

True enough, the river's perennial surface flow has mostly disappeared in the twentieth century—a lessening stream. But to describe this circumstance in moral terms of decline would be ahistorical and a disservice to the discipline of environmental history. Environmental history has more to offer than simply an ever-lengthening catalogue of human-caused ecological disasters. Neither does the river's history express a Whiggish faith in pursuit-of-happiness progress such as a modernist might espouse. One set of narrative-driving assumptions is as potentially myopic as the other. Likewise, each category of perception about the river must be understood both for its insight as well as its limits of vision. The archaic vision of the river might speak of the water's painful suffering: the river groaned as it declined, or gurgled with happiness at moments of increase. The water in the channel felt sadness when sinking beneath the sand, and jealously resisted the excitation of its molecules during the process of evaporation. A modernist would tend to describe the water in the river with no such anthropocentric terms, but would apply the moral language to the society itself. The human cultures on the river's banks would have experienced painful decline and prideful success, tearful sadness and giggling joy. I agree with the postmodern perception in that linguistic constructions will continue to describe the river and define its meaning for human society. I remain a materialist, however, convinced that just as a river existed before human cultures arrived in the valley, before my great-grandparents arrived in the valley, a river will no doubt continue to exist in one manifestation or another long after the last human culture, and I, have passed from the scene.

PART I

Archaic

I

The River as It Was
Precambrian to Late Pleistocene

A DESCRIPTION OF THE SANTA CRUZ River as it must have appeared to the first humans who beheld it requires a fair degree of historical imagination. No photographs, drawings, cave paintings or lyrical poems guide the way or give a clue to the valley's physical appearance when the first Paleo-Indians entered the region about 10,000 years ago. To whom can we turn for help in describing nonhuman, or prehuman, landscapes? Geologists understand that mountains vary in age, and rivers ebb and flow, but their concept of time, properly modified into "geologic time," holds little meaning for traditional scholars of human society. Yet environmental history seeks to expand common scholarly boundaries. To understand the human interaction with nature requires both a grasp of human society and an understanding of the prehuman natural world. While I am not proposing a strict dichotomy of human separateness from nature, the natural world has changed over time regardless of human machinations and meddling. Surely human interaction with nature has affected changes in the natural world, at times accelerating or delaying natural processes. But to thoroughly understand the human relationship with nature, the processes of change in the nonhuman world must be understood to establish a sort of baseline data point from which to measure the impact of human society. The story of a river's changing circumstances, and the effort to describe its appearance to the first human inhabitants of the valley, thus begins with an effort to explain the river's origins. The river begins in the mountains, and so to understand the river you must first understand the mountains that gave rise to the river.

When did the mountains form and the river begin to flow? The sequencing of orogenic, volcanic, and hydrologic episodes in the region allows us to posit somewhat reliable answers to these basic questions. Before there were mountains and a river in the region there was a great incursion of sea with huge tropical swamps on its southern shore. This describes the terrain in central and southern Arizona about 100 million years ago, when dinosaurs still roamed the earth. The sea was to the north and west and swamps covered the land in central Arizona. The rise of mountains to the west began about sixty million years ago; mountains so high—eventually called the Sierra Nevadas—that they blocked the rain-producing moisture from the Pacific from reaching the interior basins. Over the next ten million years the basin and range topography of the American Southwest began to form. Within the now landlocked basins mountains rose slowly, over millions of years, to thousands of feet in height. As the mountains grew, the troughs between them filled with alluvium, "older alluvium," (rocks, gravel, and other debris of erosion) forming independent basins that now lie far beneath the current surface of the river valley. Finally, about a half-million years ago, when sediment had filled these sub-basins with alluvium to the point of overflowing, a river perhaps resembling the current Santa Cruz appeared and began flowing through the newly formed valley, transporting for the first time sediment, "younger alluvium," between the formerly independent basins. The configuration of these long-since-buried sub-basins is uncertain, but the geological record nonetheless provides evidence of the pace and nature of these basic topographical features.[1]

The mountains that surround and give rise to the present Santa Cruz River began their formation about fifty million years ago—rather recently in geologic time. The oldest era in geologic time is the "Precambrian," over 600 million years ago, and the mountains that surround the Santa Cruz River valley contain lithologic formations—rocks—native to this period.[2] However, the ground movement that resulted in mountain building in the region occurred more recently, in "low-angle detachment faults" as part of "metamorphic core complexes." These phrases express a clear meaning to geologists, but may sound like gobbledygook to others. Suffice it to say, the first mountains, or hills, in the region, began as gently sloping bulges in the earth's surface that eventually snapped and rose as low-standing ridges. Subsequent to the first uplifting, shallow basins formed between the blocks of rock. In this first manifestation of the basin and range topography, the ridges stood only about 2,000 feet in height. This early geology of basins and ridges now rests from 2,000 feet to 20,000 feet below the current surface of the Santa Cruz River valley.[3]

Shallow basins and low-standing ridges gave way to the more recognizable basin and range topography through a recurring, long-evolving series of tectonic and volcanic events beginning about thirteen million years ago.

The Santa Cruz River flows through the basin and range terrain of southern Arizona. The mountains rise to elevations of more than 10,000 feet, trending from the southeast to the northwest. The river flows generally from south to north except for its characteristic about-face from a southward to northward flow around the Patagonia Mountains.

First came the high-angle faulting that created the mountains as they more-or-less appear today.[4] These high-angle slopes shot up craggy and sharp, abrupt slices of rock jutting upward, with broken pieces of mountain sliding down the steep slopes. As the mountains rose through the uplift of block and thrust faulting, the intermontane areas sank and receded. During periods of tectonic peace, sediments from the arisen mountains eroded down into the basins. Periodically throughout this process of uplift and infill, volcanic extrusions occurred, most recently about one million years ago, forming Sentinel Peak ("A" Mountain), Tumamoc Hill, Martinez Hill, and the Black Mountains.[5] Geologists debate the ending point of the tectonic activity that formed the basin and range geomorphic province. Some maintain that the "main phase" of tectonic activity ended six to three million years ago, while others maintain the activity continues into the present.[6]

The resulting basin and range landscape came to encompass two regional configurations, the Sonoran area and the Mexican Highland section, as labeled by the geologist Nevin Fenneman in 1931. Fenneman described both sections as "approximately half mountains and half plains" trending toward the northwest.[7] The Santa Cruz River valley from Tucson south lies in the Mexican Highland subprovince, while the Avra Valley occupies a position north and west of Tucson on the topographical boundary between the two regions. The Sonoran region, extending from the Santa Cruz Flats to the Phoenix Basin, is lower in elevation and somewhat drier, with resulting variations in vegetation. Conversely, the Mexican Highland section is higher in elevation, somewhat wetter, with fewer cacti species and more grasses and woody shrubs.

The bedrock that anchors the mountains surrounding the river serves as an "impermeable and nonwater-bearing basement formation" that slopes down beneath the valley.[8] Closer to the mountains bedrock formations are nearer to the surface, but in the central valley bedrock lies far beneath the desert floor. On top of this base lies the oldest alluvium—the sediment, volcanic detritus, and general lithological debris from the surrounding mountains, sometimes red in color due to the presence of iron oxide. In the Tucson Basin over 20,000 feet of gravel, sand, and silt have been eroded down from the surrounding mountains.[9] To the south in the upper valley, the Nogales Formation was deposited thirteen to twenty-five million years ago, and consists "predominantly of volcanic detritus," to a depth of 1,000 feet.[10] Near the San Cayetano Mountains alluvium occurs to a depth of 4,300 feet.[11] These batches of "older alluvium" entered and filled the shallow sub-basins long before the proto–Santa Cruz River appeared.

Younger alluvium covers the older sediments, and was deposited by the river that first appeared in the middle Pleistocene, 0.7 to 0.25 million years ago, or to split the difference, a half-million years ago. Initially, alluvium washed down from the mountains into independent playa lake basins (temporary, shallow lakes forming from rain and runoff, then evaporating during dry spells). Not until a through-flowing river appeared did alluvium spread between the isolated basins.[12] These more recent deposits, called "surficial" by geologists, include gravel from as early as one million years ago, and all the debris left by later floods, erosion, and wildcat trash dumps "associated with modern fluvial systems."[13] The record of this proto–Santa Cruz River is clearer in the middle basin near Tucson, where arroyoization—a deepening of the river channel during flood episodes—has exposed much of the recent alluvium. In the upper basin the picture is less clear. Hydrologist David Helmick posits that the proto–Santa Cruz River may merely have been smaller at these higher elevations or perhaps even flowed steadily south.[14]

Erosion of the more recent basin fill, interspersed with redeposition, has formed the terraces that slope down toward the river. During the last 10,000 years a period of redeposition has left 100 to 150 feet of "unconsolidated fill" in the central valley. This represents the modern floodplain, which is being cut by the present-day arroyo. One advantage of the river's arroyoization is that the exposed banks to a maximum depth of forty feet display the last 8,000 years of the river's alluvial history, which coincides, quite handily, with the arrival of human society in the river valley.[15]

Hydrologists refer to the bedrock as impermeable to water. The older alluvium has low permeability, due to the high clay content and prevalence of "consolidated rock." The younger alluvium is generally coarser, and enjoys high permeability. In general the surface flow of the river occurs where the water table has risen sufficiently in the younger alluvium to support it, or where an impermeable formation of bedrock lies close enough to the surface to force subsurface water aboveground.[16]

Replenishment of water into the older alluvium occurs as moisture from rain and snow on the mountains trickles down through cracks and fissures until it reaches the junction between the "hydrologic bedrock" beneath the mountain and the older alluvium. Water then seeps down the underground slope toward the center of the valley. Hydrologists believe that at this point, some water is passed from the older alluvium to the younger alluvium above it. Although the permeability of the older alluvium is limited, the total volume of water present is quite large due to the huge volume of alluvium itself. Hundreds of millions of acre-feet of water (an acre-foot is the amount of water it would take to cover an acre of land with water one foot deep, or 325,851 gallons) rests at present in the older alluvium, but this water is generally inaccessible below depths of about 1,000 feet.[17]

The highly permeable younger alluvium allows water to move freely, albeit slowly, a little more than 200 feet per year.[18] Recharge occurs from above and below. While a small amount of water percolates up from the older alluvium, most of the recharge comes from above, through direct infiltration from precipitation, seepage from irrigation and floods, and any other source that augments the surface flow of the river, including modern-day discharge from sewage treatment plants, and even water from far beyond the watershed delivered in the Central Arizona Project canal. In a sense, the younger alluvium acts like a giant sponge, or in more technical terms, a "line sink/source." If the alluvium is saturated, then it delivers water to the river channel. Conversely, if the alluvium is dry, it draws water from the river channel and the surface flow disappears into the sand.[19] Over the last one hundred years or so the water table has fallen to such an extent over so much of the river's course that the river channel and younger alluvium today acts primarily as a line "sink" rather than "source." It is as if the top of the sponge is now mostly dry, and so water added to the formation must travel down to replenish the interior of the sponge before allowing the surface to glisten once again with moisture. Prior to extensive groundwater pumping in the twentieth century, the younger alluvium in the Tucson Basin—a more-or-less saturated sponge—held about seventy million acre-feet of water above depths of 1,200 feet.[20]

Just how much water is seventy million acre-feet? I suspect that such a statistic does not communicate much useful information, nor, in fact, does the basic unit designation of one acre-foot, since I presume many readers have only a fuzzy understanding of the size of one acre. Here then is a variation on the acre-foot designation: one acre-foot of water would cover a football field, including both end zones, nine inches deep. As to the vast quantity of seventy million acre-feet, our flooded football field, if somehow the water could be confined to that surface area at its base, would constitute a pillar of water 10,000 miles high.

Climatological variations also affected the river's circumstances 10,000 years ago. Paleontologists may argue over the exact pattern of climatological change in the region, but agree that rainfall and temperature patterns fluctuate as part of natural cycles. The development of weather systems in the Pacific and Caribbean affect the amount and sequencing of rainfall events and temperature variations in the Southwest. For example, Julio Betancourt and other scholars posit that shifts in the "El Niño" system in the Pacific historically have exercised an episodic influence over weather patterns in the Santa Cruz River valley, resulting in periodic floods and arroyoization in the river channel.[21]

The debate over the sequence and extent of climate change takes place within a range of consensus. Most scholars agree that semiarid conditions in the region have prevailed for the last 6,000 to 8,000 years. Earlier climate patterns were probably wetter, with heavier winter precipitation resulting in more woodland trees and shrubs, such as pinyon pine and juniper. The debate centers around when this wetter pattern changed to dryer circumstances.[22] No doubt the river changed as a consequence of the changing weather patterns. The wetter conditions 10,000 years ago probably created a broad, shallow, meandering or braided stream that must have been quite alluring to the first human arrivals.

It was an enticing scene, but not a static one. Geologic realities coupled with climatological variations have, over time, created natural pressures and influences causing the river to meander, flood, gouge channels, slow to form marshes and swamps (*cienegas* in the Spanish), dry up, change course, and deposit fresh alluvium. It is clear in the amount and distribution of younger alluvium that the river has a long history of transporting sediments and detritus. It is also clear in those areas where recent erosion has exposed the river channel that the river has also undergone periods of flooding and arroyoization.

The earliest history of the river is buried under hundreds of feet of younger alluvium. Drill holes and magnetic soundings can tell us only so much, or at times tease us with possibilities, such as Helmick's south-flowing hypothesis. What is clear, however, is the long history of deposition and the relative certainty that periods of erosion and paleo-arroyoization have interrupted the calm history of a steadily flowing stream. It is also important to understand that the river may have always manifested itself in different configurations along its entire course. This seems a safe assumption, beyond simply the commonsensical, given the wide variety of topographies and vegetation zones through which the river has always flowed. These variables of topography and vegetation have affected the river's flow, just as overriding patterns of geology, hydrology, and climate have impacted the river. Over the past half-million years the river has meandered, flooded, and dried up according to the force and power of the variables acting upon it. In its most recent history, the human variable appeared in the valley. The first humans traveled to the valley because of the river and ecosystem it supported. Grassy plains and spruce-covered slopes provided fodder for large numbers of grazing mammals. The first humans in the valley were Paleo-Indian big-game hunters, and they had happened upon a hunter's paradise.[23]

What did these nomadic hunters see as they entered the valley? We might reasonably assume the hunters, traveling in small groups organized around family units, entered the region from the west and north, following the common explorer's stratagem of traveling upstream along rivers and their tributaries. In this scenario, the hunters would have followed the Gila River upstream to its confluence with the Santa Cruz River. Who knows what they called these rivers? Perhaps the two in comparison would have been simply "Big River" and "Little River." Whatever the name, the Gila, or Big River, flowed constantly with an impressive current that in places was no doubt treacherous and impossible to cross. If they arrived in spring, runoff from snowcapped mountains would have made the river a torrent. Perhaps it is safer to assume an early autumn arrival, with summer rains slackening, and nature for a time quiescent. Bear grass and other perennial and annual varieties of grass, interspersed with Joshua trees and yucca cactus, filled the plains surrounding the Big River in its lower regions, the grass becoming more plentiful and lush and the cactus diminishing as the elevation steadily heightened upstream. In the higher elevations near the confluence of the Big and Little Rivers, many species of large grazing mammals—ground sloths, bison, camels, horses, and mammoths—ignorant of human predators, grazed on the thick and bountiful grass, making the area a prime hunting ground for the newcomers. Oak woodlands, along with pinyon and juniper stands, occurred down to elevations of 1,800 feet, but the plains remained mostly grassy given the prevalence of wildfires, which replenished the grass while destroying brushy seedlings. Hills and mountains in the distance marked the extent of the valleys. The lower volcanic hills, rocky and abrupt, showed grass and the occasional stubborn juniper growing between the craggy debris of rocks and boulders. The higher, older mountain peaks, covered with snow through most of the year, shimmered blue in the distance with thick stands of spruce, fir, and mountain juniper. Pinyon and juniper descended the slopes of the mountains onto the hills lining the valley.[24]

The Little River joined the Big River as little more than a stream after traveling through a broad, flat grassland that would take perhaps three days to cross on foot. The prairie extending to the south, interrupted by the occasional volcanic outcropping, promised more good hunting. Animals proved especially vulnerable in those places where the gradually drying climate caused small watercourses, like the Little River in its lowest reaches, to reduce itself to springs and seeps. As animals congregated around these water holes, hunters found easy prey. Perhaps the hunters considered themselves lucky, or thought of the land as welcoming, given the blessing of bounty. Traveling south and east, the hunters may have been struck by the odd-looking peak, later called Picacho, and, drawn to the springs at its base, may have encountered a plethora of indigenous wildlife.

The Little River increased in steady flow beyond the peak as the hunters neared the middle basin of the valley, where higher mountains narrowed to pinch the river as it coursed into the flats the hunters had just traversed. Conifer forests were clearly visible on the 9,000-foot peaks, and the pinyon and juniper stands encroached on the grasslands at the base of the mountains. Cottonwood and sycamore trees lined the river over the last reaches of the flats, growing thicker beyond the narrows into the middle basin. A major tributary, later called Rillito, joined the river just beyond the narrows to the south, and another promising landscape stretched to the east in the valley. More oak woodlands and juniper stands choked the foothills to the north of this second little river. The first Little River meandered through the western side of the basin, feeding from seeps and springs from low volcanic mountains to the west. Cottonwood and sycamore towered over the river, and interspersed among the trees in several marshy areas were thick stands of willow, elderberries, ash, walnut, hackberry, and catclaw.

Rainfall was plentiful in the river valley, falling in regular seasonal patterns, although the long-term trend, unbeknownst to these first wandering hunters, was toward dryer and warmer conditions. Also unbeknownst to the Paleo-Indians, the rainfall that fell on the watershed was slowly making its way into the river's aquifer in the younger alluvium. There it would reside undisturbed except for a slow creep toward the northwest, traveling in 10,000 years more than thirty miles underground from the place where it first seeped into the aquifer. This ancient rainfall is now being pumped by city water departments and ranchers throughout the valley.[25]

Recent volcanic mountains once again narrowed the river's channel at the southern end of the middle basin. Grass attempted to cover the hills, but the coarse black rocks showed through clearly. No cactus grew on the slopes of the hills at these elevations, where cool temperatures and moist conditions limited the range of saguaro and other cacti species. Beyond these volcanic narrows the valley opened to an enticing view of more grassland surrounding a clearly marked river channel. Springs just south of the narrows created another marsh of willow, hackberry, and catclaw, and furnished much of the flow through the middle basin, but south of the marsh the stream diminished. Nonetheless, eons of steady flow and saturation of the alluvium to depths of hundreds and thousands of feet had created an aquifer that could support cottonwood and other thirsty plants even where the stream's flow trickled. More tall mountains (Santa Ritas) with twin peaks 10,000 feet high bordered the valley to the east, and the cooler temperatures in these higher elevations kept snow on the peaks year-round. Conifer forests spread far down the sides of the mountains, and juniper trees spilled onto the foothills. Further up the valley as terraces lined the river, juniper and pinyon pine found footholds closer to the river, restricting the grass to the flatter mesas and plains. Wildlife may have taken refuge in these canyons cloaked with growth, and the hunters may have lost some game in these stretches where wounded critters could scurry away to hide and die alone.[26]

In the terrace and mesa country side canyons proliferated, each with a stream tumbling down the pebbled channel from the low mountains to the west, and from another range of camelback mountains (San Cayetanos) to the east. Just beyond the camelback mountains another major tributary (Sonoita Creek) joined the Little River from the east. Temperatures were cooler in the valley, and colder still in these higher elevations, and pinyon and juniper stands dominated the landscape. The river began twisting and turning through the broken country, but the hunters may have found congenial campsites among groves of walnut trees in one particular south-trending canyon (Nogales Wash). As the river turned first east, then south, then east again, and finally north, the hunters must have realized that an unusual landscape confronted them. Likely not many other rivers in their collective experience had made such an extreme change in direction. Entering the high valley (San Rafael) that spawned the river, bordered by more 10,000-foot mountains to the east, the hunters traversed the last stretch of the river to its headwaters in low hills at the valley's northern limit. Pinyon and juniper covered the basin of the valley, and the first snowfall may have recently dusted the ground. The higher elevations with regular snowfall and freezing low temperatures through much of the year limited varieties and prospects of grasses. But even here forests held game, and also dangers, as saber-toothed tigers lurked in the caves in the hills.

If they had been intent on making just a quick survey of the river and its valley, the journey might have taken the hunters on a walk lasting half a moon's cycle, or about two weeks. But who knows how long it took them to travel the entire course of the valley, and whether they did so with any sense of wonder or amazement. What is clear, no matter what appraisal the first humans made, is that the valley was lush with plant life, wildlife, and water. Cooler and wetter conditions, replete with late Pleistocene large mammal species such as mammoths and ground sloths, made the valley an enticing site for human incursion.

As the last ice age ended at the close of the Pleistocene period, warmer and dryer conditions gradually prevailed in the valley. The large mammal species disappeared, perhaps hunted to extinction, or perhaps a combination of hunting and changing environmental conditions simply caused the species to migrate out of the region. Paul Martin's theory of "Pleistocene Overkill" posits that the newly arrived Paleo-Indian populations, who recently crossed over the land bridge between Asia and North America, over-hunted to the point of extinction as much as 70 percent of the large mammal species at the end of the Pleistocene. Martin pointed to the fossil evidence of mass extinctions and the number of fossil remains including stone spear points; however, clouding the issue of "overkill" is the roughly concurrent change in the environment. The increasing aridity in the Southwest may have contributed to the extinctions or migrations of large mammal species from the region.[27] Since the first people in the region were hunters, it is safe to assume that as game species migrated or died off, the people moved on, either following the diminishing herds out of the region, or moving on to other hunting grounds once the valley had been denuded of prey. Afterwards, a period of humanless quiet descended on the valley. The absence of human voices was brief, however, and it is clear that the first humans in the valley soon were followed by other groups of human beings who stayed longer and found ways to extract subsistence, and even surplus, from the increasingly dry environment. These people stayed put and the archaeological record becomes deeper.

The first migratory humans in the valley left no trace of habitation sites and obviously placed no demand on the river that could be recorded or measured. At the time the Paleo-Indians arrived, the wetter climate contributed to a more substantial flow in the river. The dominant factors in the river's circumstances were climate and geology. Spring thaws from mountain snowpacks might have caused routine floods in the valley, while cooler overall temperatures probably kept the river flowing through most of its course throughout the year. In addition to these general influences, specific factors such as overgrazing of the watershed by browsing wildlife also could have affected the river. Although predators would have generally held the grazing populations in check, had that balance ever been lost, overgrazing might have occurred. Lightning-caused grass fires might also have occasionally denuded the watershed of grass cover. Simply, or in combination, these factors might have caused summer or autumn floods on the river that cut arroyos or furthered the spread of the floodplain beneath the terraced slopes. But the most dominant factors affecting the river remained climate and geology. Any arroyo cutting would have faced routine spring aggradation as the heavier flows readily transported sediment down the river valley. As to the human factor in this scenario, the Paleo-Indian hunters were superpredators and their presence in the valley would have ended for a time the possibility that any browsing mammal species might overgraze the watershed. During the late Pleistocene over 100 extinctions occurred, leaving many vacancies within the grassland ecosystem. For example, Pleistocene horses became extinct during this period, vacating a niche that remained unfilled until Spanish horses and donkeys arrived in the sixteenth century. Feral populations of these European horses and donkeys soon reoccupied the niche that had been vacated 10,000 years earlier. Following this line of reasoning, the arrival of humans to the valley may have coincided with thickening grass cover on the watershed and a declining incidence of arroyo-cutting floods.

With the transition to a dryer climate at the end of the Pleistocene, the changing pattern of slow meander, cienega, dry stretches, and periodic flooding and arroyoization took hold in the valley. Into this setting the Clovis people emerged, and the narrative of human presence in the valley intensified.

From Clovis Points to Maize

10,000 B.C.—A.D. 1500

ANTHROPOLOGISTS DESCRIBE AN ARCHAIC perception of nature, and rivers, held by the prehistoric Indians who occupied the Santa Cruz River valley. An animistic view is evident from the material remains of the sedentary cultures of the region, but what surprises is the level of technical expertise some of the prehistoric Indians brought to bear. Their engineering skills foreshadowed modern techniques for surface water manipulation, and one must wonder if these prehistoric peoples adopted a less mystical or spiritual perception of the river than scholars have presumed. No documents or material evidence speak to this question of nuanced perceptions, but the effects of human manipulation of the river become so pronounced during the prehistoric period that theories about the Indians' motivations and perceptions might be extrapolated from the engineering works themselves. Regardless of motive or vision, the prehistoric residents of the valley were the first to affect the river with a force rivalling both climate and geology.

A population explosion of sorts occurred about 9500–9000 B.C., as the Clovis complex of Paleo-Indians found new ways to utilize the area's resources. Archaeologists have found many Clovis points—sharpened stone-cutting tools and projectile points—in the San Pedro River valley to the east, though only a few have been found in the Santa Cruz River valley, probably because most of the late Pleistocene to early Holocene deposits along the Santa Cruz River remain covered by alluvium. Likewise, the paucity of Paleo-Indian deposits upstream from the Tucson Basin is probably due to their location several meters beneath the surface. Although two sites uncovered during highway construction near Tumacacori give evidence of human inhabitance from about 2000 B.C., five thousand years removed from the Clovis culture, little is known about the extent and nature of Clovis settlement in the Santa Cruz valley.[1] As the climate of the region became drier in the early Holocene, the Clovis culture gave way to Folsom and Plano groups who followed the large mammals surviving Pleistocene Overkill into the short-grass regions of the central plains. Archaic hunter-gatherers (here "Archaic" refers to a time period designation used by anthropologists) succeeded the Paleo-Indian big-game hunters, pursuing lifestyles that included hunting and other subsistence patterns more amenable to changing environmental conditions. The Archaic peoples were small in number and were restricted to small riparian areas during the periods of drier climate, 5500 to 3000 B.C. During wetter times around 3000 B.C., the Archaic people increased in population as their utilization of the bioregion's resources became ever more efficient. Sedentary cultures developed as subsistence patterns evolved from plant gathering and foraging to "incipient agriculture," (the earliest forms of cultivation along streambeds and other naturally conducive terrain), which appeared around 2000 to 1500 B.C.[2]

Variations among these Archaic cultures are sometimes hard to define. Archaeologists characterize the different groups on the basis of the tools they used, but the many "cultural cross-currents" resulting in part from nascent trading networks cause arguments about distinctions and periods. In general, Archaic sites occur near cienega deposits, where plant and wildlife species would have occurred in greater abundance. At least eighteen Archaic sites have been identified near the floodplain of the Santa Cruz River and its tributaries in the Tucson Basin. Archaeologists found one Archaic site in Pantano Wash under fifteen feet of alluvium, with indications of habitation extending from 3000 B.C. to A.D. 100–200. The site was significant for the presence of *zea*, an early form of corn, which indicates a high degree of sedentism. Another site near Martinez Hill gives a geologic picture of the human presence in the valley. Archaic deposits at the site occur eighteen to twenty-one feet beneath the current surface of the valley. Fifteen feet of alluvium separate the Archaic deposits (some carbon-dated to 3900 B.C.) from more recent Hohokam deposits resting three to five feet beneath the surface.[3] More than 3000 years passed during the deposition of the fifteen feet of alluvium separating the early Archaic sites from those of the Hohokam. This is not to say a void of human habitation occurred during the interval between archeological sites. Rather, a period of slow cultural development took place, with the resulting transitions and linkages between earlier and later peoples unclear to archaeologists. Significant for this narrative is not so much the continuity of human

Santa Cruz River Arroyo in the Tucson Basin

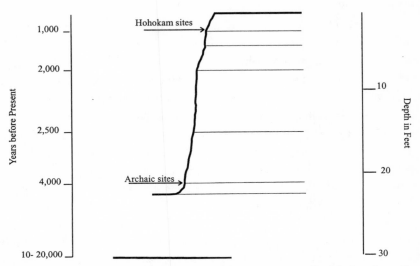

This diagram of the arroyo in the Tucson Basin shows the relative depth below the surface for Hohokam and Archaic sites. The arroyo varies in depth as low-flow conditions add sediment to the channel floor and flooding episodes deepen the channel by transporting sediment downstream.

presence in the valley, but the development over time of a culture, the Hohokam, that came to exert a clear influence on the diminutive river.

Archaeologists debate the origins and chronology of the development of irrigated agriculture and the Hohokam culture in the desert Southwest. Some scholars posit indigenous evolution from earlier cultures to explain the Hohokam, while others suggest influences from Mesoamerican migration. Current scholarship describes the Hohokam culture as extending roughly from A.D. 650 to 1450, with earlier progenitor cultures occupying the Early Agriculture Period, from 1200 B.C. to A.D. 150, and the Early Ceramic Period from A.D. 150 to 650. Habitation sites with evidence of irrigated agriculture in the Tucson Basin appear as early as 1000 B.C., but these inhabitants of the valley bear uncertain relationships with the Hohokam. Archaeologists draw distinctions between these cultures from various pottery types and living patterns, and most important, from the extent and sophistication of water resource manipulation. The peak of Hohokam culture occurred from A.D. 900 to 1300. By A.D. 1500, areas inhabited for thousands of years along the Santa Cruz and surrounding rivers had been abandoned, with tens of thousands of acres of land, formerly irrigated, now dry and fallow.[4]

The word *Hohokam* means "all used up" in the language of the Pima Indians, who themselves may have been descendants of the Hohokam.[5] Among archaeologists the Hohokam are renowned for their canal building and use of irrigated agriculture, traits as distinctive in their own way as the cliff dwellings and pueblos of the Anasazi, northern contemporaries of the Hohokam. The "all used up" of the Pima Indians seems a clear reference to the Hohokam's disappearance from the archaeological record around A.D. 1450 to 1500, and may be reference to the depletion of water and other resources crucial to Hohokam survival, perhaps even to population levels growing beyond the ability of the ecosystem to support them. The question of the Hohokam's possible demise due to overpopulation vexes archaeologists, who tend to shy away from such finite pronouncements. After all, how large a population is too large? Nonetheless, the pressures and consequences of increasing population density is a central question in the Hohokam's history, and relates directly to the degree and effectiveness of their manipulation of the river and the semiarid environment.

Population levels of the Hohokam in the region reached densities unmatched until the twentieth century. Usually scholars speak of these peaks of population and culture only in relative terms, and one might ask how an assertion can be made that heightened population density reflected cultural apogee.[6] Can we even be certain that an increasing population was valued by the Hohokam, or that since the Hohokam population increased, it did so as the result of deliberation, planning, or debate? Perhaps the most prominent early scholar of the Hohokam, Emil Haury, speculated that the simple and unchanging form of Hohokam human figurines indicates a pro-growth ethos: "I suggest they [the figurines] were tangible objects associated with a system of worship, or a means of supplication, which affected everyone in the culture. Most likely they were a form of house blessing, a means of insuring increase, of the family, of crops, and through these, securing the fulfillment of the society's needs. A deep-seated conviction of their efficacy may well have restrained morphological changes over so long a period of time." It would seem reasonable to assume that the Hohokam strove to maximize the effectiveness of their subsistence practices, if not to increase their own numbers, then at least to make more secure their current circumstances in the semiarid region.[7]

Precipitation rates within the bioregion vary, from over twenty inches per year in the mountains to five inches per year in the central desert. The Tucson Basin averages from eleven to twelve inches of precipitation per year, with evaporation rates in the glaring sunshine at times amounting to seventy-seven inches per year, or seven times the average rainfall.[8] Another quality of the precipitation in the area is its extreme variability from year to year. Variations between wet and dry years—outside the realm of drought cycles—may amount to 55 percent, often determining the success or failure of agricultural pursuits. As previously mentioned, these semiarid conditions have prevailed in the region for about 8,000 years, and vegetation throughout most of the Hohokam domain was characterized by mesquite and palo verde types of plant life, similar to the types of vegetation seen in the region today. It is fair enough to call the Hohokam "desert dwellers," as long as readers unfamiliar with the terrain keep in mind that rivers flowed through the Hohokam's land, and native vegetation and wildlife provided bountiful sustenance.[9]

Given the semiarid conditions in the region, the proliferation of the Hohokam culture was predicated on irrigated agriculture. The Hohokam drew water into their canals from rivers with long reaches into the surrounding mountains. Little more than shallow ditches at first, the canal systems grew and developed with the culture. In the center of Hohokam settlements, along the Gila and Salt Rivers in central Arizona, Hohokam canals reached a cumulative length of 180 miles during the Classic Period, about A.D. 1300. This marked the peak of a contemporaneous network. Haury and others have documented an additional 120 miles of canals in the Phoenix Basin, but these were earlier Hohokam constructions and not part of the Classic Period system.[10] At its peak, archaeologists estimate irrigated acreage in the Phoenix Basin to be 30,000 to 60,000 acres. This extensive system of canals in the Phoenix Basin has no exact corollary in the Tucson Basin. Perhaps a smaller population in the southern region could not support such extensive works as in the Phoenix Basin, or perhaps the smaller flow in the Santa Cruz River made such massive works unrealistic. The Santa Cruz simply could not provide enough water to fill larger structures (a nineteenth-century engineer measured the Santa Cruz River's "low flow" volume to be 1,000 miner's inches, or about one acre-foot every forty-three minutes). Nonetheless, the smaller canals and other manipulative structures found in the Tucson Basin exhibited the same engineering skill as the larger systems in the center of the Hohokam settlements.[11]

When viewing excavations of the canals, one finds it difficult to imagine that the Hohokam accomplished engineering works of such skill and complexity using only neolithic implements of stone and wood. Moreover, the configuration of the canals shows a keen awareness of topography and hydrology. The canals in the Phoenix Basin were immense in size, ten feet deep in places, and thirty feet wide. The Indians lined the canals with a sort of adobe plaster so as to limit the loss of water through seepage (a practice ignored by modern engineers in the 1930s in the initial construction of the All-American Canal in southern California). The slope and gradient of the Hohokam canals also show remarkable engineering skill. The Indians placed the diversion headgates far enough upstream to reach fields on terraces downstream overlooking the river. The gradients allowed water to flow swiftly enough to limit evaporation in the harsh desert sun, while not flowing so quickly as to cause erosion within the canals themselves.[12]

In addition to the large canal systems, the Hohokam employed smaller systems and less-sophisticated methods of increasing soil moisture. Many of these strategies are still employed by local native populations, such as the Tohono O'odham in the Santa Cruz River valley. The aim of these manipulations of terrain was to increase runoff from normal precipitation and then channel it into areas under cultivation. To achieve this, the Hohokam cleared vegetation on slopes upstream from their fields and used rocks and brush to create "diversion weirs" to direct runoff water onto the desired locations. The natives also used living plants, such as cottonwood and willow, to devise "fencerow silt traps" to catch runoff moisture and silt. Tohono O'odham farmers in southern Arizona still use a similar system of mesquite posts interwoven with brush "as diversion weirs . . . water spreaders and baffles."[13]

Environmental factors may have also played a role in the Hohokam peak in population around A.D. 1000. Dendrochronologists describe a period of increased precipitation for the Southwest that seems to coincide with the peak in Hohokam population. During this period the natives reached their maximum population by the spread of agriculture beyond the close proximity to rivers.[14]

Typical rainfall of eleven inches per year in the valleys and floodplains of the bioregion—even if augmented by unusually high precipitation rates around A.D. 1000—would not have been sufficient on its own to support most crops. Two factors combined to allow the Hohokam to exploit non-riverine environments in the semiarid environment. Irrigation techniques and the modification of local terrain to augment the amount of runoff moisture delivered to their crops was the first. The second factor was the introduction of new drought-resistant crops (probably through Meso-american contacts), such as onaveno maize and tepary beans.[15]

Although known for their engineering works and manipulation of the semiarid environment, the Hohokam never turned away from reliance on native and indigenous resources. Three general environments in the Hohokam's territory helped provide for their subsistence. In the "upland zone" on the higher slopes of the nearby mountains native populations harvested fruit from several varieties of cacti (saguaro, cholla, prickly pear, and barrel), and hunted game animals such as desert bighorn sheep, mule deer, and jackrabbits. In the "bajada zone" on the terraces and mesas nearer the river, "desert riparian" resources such as mesquite beans were available. It was on these slopes that manipulative structures were best deployed, and where semi-domesticated plants such as agave were grown. One bajada slope under cultivation covered almost 1,200 acres and included over seventy-two miles of terraces and check-dams. In the "floodplain zone" of the river and tributary washes mesquite forests flourished, as did cotton-wood trees and desert willow, with grasses and desert shrubs generally interspersed. The wildlife species found in the river environs, especially within the dense stands of mesquite, included coyote, bobcat, raccoon, grey fox, cottontail and jackrabbit, various rodents, Gambel's quail, road-runner, and the expected birds of prey. Most significant for the prehistoric farmers, the floodplain zone provided fertile alluvial soil as much as twenty feet deep, without the troublesome southwestern hard-packed *caliche*.[16]

The water table was shallow enough to be within easy reach of crop root systems, and the growing season in the semiarid region was 245 days, which allowed the planting and harvesting of at least two crops annually. The first crop was probably planted in November and harvested in May. A second would be planted in June and harvested in October. Alternatively, two crops of frost-sensitive corn could be planted, the first in April, harvested in June, and again in July, harvested in early October. The shallow water table also allowed for wells, six to eight feet deep, outside the houses of the Indians. And the river itself provided fish for the Indians to eat.[17]

The centrality of the river to the subsistence practices of the Hohokam neither surprises nor perplexes, but the presence of wells outside the pre-historic dwellings raises some interesting questions. Did the Hohokam, master engineers of surface irrigation systems, also possess an awareness and knowledge of the aquifer's nature? Could the Hohokam have brought a modern perception of the river to the valley 500 years before industrial society arrived? Their manipulation of the river's surface flow clearly approached the level of intrusiveness brought by modern engineers, but a modern perspective encompassing an image of the immense water supplies hundreds of feet beneath the surface of the valley must have eluded the Hohokam. Their shallow wells into the floodplain were simply a natural corollary to the headgates and canals which harnessed the precious surface flow. The shallow water table could be explained as simply the gritty oozing cousin of the clear-running stream. We can easily imagine a supplication to watery spirits when the Hohokam drew water from a well. Their cosmological explanations for the origins of this subterranean life force would have been animistic, mystical, and archaic, no matter how sophisticated their technical expertise.

On the other hand, in their effort to maximize use of the river's surface scarcity, the Hohokam may have exerted an influence so profound that they changed the river's course through the middle basin. Geohydrologists suggest that the river's historic path—from recorded time through the early-twentieth century—through the Tucson Basin resulted from a prehistoric effort to redirect the river's flow to Hohokam agricultural fields. According to this theory, the "natural" course of the river followed a path a mile or two east of the river's historic channel. To the researchers, the course of the historic river hugging the western edge of the basin does not make geologic sense. Rather, the so-called spring branch of the river, east of the historic main stem, follows a more reasonable course through the basin. The theory holds that the Hohokam dug a headcut—an excavation in the river channel meant to intersect the water table and redirect the flow of water into irrigation canals—just south of Martinez Hill, perhaps at the outlet of the cienega where the surface flow of the river returned after its subterranean traverse of the valley from the area of Tubac. The headcut was intended to increase and regularize the flow to agricultural fields downstream and to the west of the existing river channel. Subsequent floods may have eroded the canal and enlarged the headcut, causing an arroyo to develop along the hand-dug ditch, from the source of the stream to the fields downstream. This entrenchment and lowering of the irrigation canal would have intercepted the surface flow at the site of the headcut, leaving the previous eastside channel high and dry. While groundwater levels would have remained high enough to create springs and seeps along the former course, ultimately resulting in the spring branch of historic times, any continuous flow would have followed the deepening irrigation canal.[18]

The ramification of this theory is significant. If the Hohokam manipulated the river to such an extent that they altered its natural course, in addition to augmenting surface flows in the groundwater through a myriad of techniques and strategies, the prehistoric Indians brought to the river a humanistic force and power not seen again for 500 years. Not until the late-nineteenth century would human engineering in the development of water resources match the Hohokam's level of intrusion.

The Hohokam no doubt utilized resources for food that later European residents would have avoided.[19] It is difficult to know exactly what the Hohokam ate, but their living sites give evidence of their dietary habits. Grease droppings from animal fat stained Hohokam oven pits, and the pits contained residue of grass species that were presumably baked as a food source. Also, we know the Hohokam cultivated beans, which seems a natural correlation with the historic Papago (Tohono O'odham), so named by the Spanish as "bean eaters."

The correlation between the eating habits of the prehistoric Hohokam and the historic Indians seems a natural one. For example, the historic native populations utilized cactus for food, and certainly the Hohokam did too. Both the historic and prehistoric Indians consumed the ripened cactus fruit, and also on occasion the plant itself (for example, *nopalitos,* or young, tender, prickly pear pads). The Hohokam apparently cultivated agave (century plant), which means they almost certainly enjoyed the common alcoholic drink, mescal. Pima Indians also made an alcoholic drink called *tesguino* from fermented corn, and "Papago wine" was made from the juice from saguaro cactus fruit.[20]

Maize (corn) was a basic staple of the Hohokam and historic Indians, and the Spanish priests and settlers who arrived later adopted many of the traditional Indian foods made from corn into their diet. The simplest form of maize, according to the Jesuit priest Ignaz Pfefferkorn, was *posole,* prepared by boiling the corn until the kernels "burst and become soft." The priest said the Indians then ate the "insipid dish . . . without further ado." *Atole,* a sort of corn-based porridge sweetened with sugar and enlivened with cinnamon when possible, was "the common breakfast of the Indians as well as those Spanish who cannot afford chocolate." *Esquita* was a form of popped corn: "the kernels bust open, and the pith breaks forth just like snow-white flowers." *Pinole* was the gruel that resulted from grinding esquita and adding water. It was often used as a convenient ration while traveling, or by soldiers in the field. Pinole was adapted to available food resources—mesquite beans, corn, and wheat—from prehistoric to historic times.[21] And, of course, cornmeal was the basis for tortillas and tamales.

Quite simply, the Hohokam exploited their environment with incredible efficiency. Scholars from Emil Haury on have recognized the Hohokam to be the most efficient culture in their utilization of the region's resources ever to inhabit the valley. This is an underlying assumption that runs through all the archaeological and anthropological studies of the culture. The scholars' admiration for the Hohokam comes in part from the recognition of the culture's successful adaptation to the region's apparently meager resources. The flip side of the assumption is that our modern industrial culture is incredibly wasteful in comparison to the prehistoric "farmers and craftsmen." We have surpassed their population maximums, but only with the assistance of our massive industrial capacity. By the time Hohokam population levels were matched in the early-twentieth century, the river valley had been linked to national and international markets, and residents of the valley relied on imports from beyond the valley for their day-to-day subsistence.

How many Hohokam derived their subsistence from the tenuous flow of desert rivers and parched surroundings? Archaeologists tend to eschew such guesswork, speaking freely of relative increases and decreases in population density, but avoiding any real numbers. But this is a basic question, and lurking just beneath the text of many studies of the Hohokam is the assumed moral: the Hohokam experience warns modern readers of the dangers of over-stretching their ecological boundaries. But if the environmental fable is to work out, then population numbers must be determined along with an understanding of the outer limits of sustainability provided by the semiarid ecosystem. Having undertaken the search for population numbers, I now better understand the archaeologists' reluctance in this regard.

The common reluctance to cite population estimates arises in part because the population ranges seem on their face to be so unreasonable. Scholars assume that the Hohokam utilized their environment most efficiently, a belief that should push the estimated population toward the higher end of any proposed range. The estimates that result are so large, however, they diverge from our commonsensical expectations.

Three methods of analysis might be used to come up with population numbers. First is the use of living space—family units per settlement—as a means of gauging overall population. Scholars have measured the spacing of pithouses (single-family living sites), and counted the number and density of living sites in a platform mound community (large village settings on raised platforms) so as to determine, rather easily, relative densities increasing or decreasing. If a reliable number of individuals per living site can be determined, then an overall estimate is within reach. The fly in the ointment is the variable age of dwellings. Archaeological digs occur on sites inhabited over long durations, with one platform mound or pithouse undergoing several, dozens, or perhaps countless, occupancies. How then to determine the number of individuals when it is unknown how many families traveled through each living space? The key issue in this method is the use-life of pithouses: the longer their use-life, the less reliable they become as measures of individuals at a particular site at a particular time. Because there is no way to be certain of the use-life, some scholars maintain that true population estimates, at least using this method, are impossible.[22]

A second method by which populations might be estimated is the analysis of labor requirements of settlements. Canals and the Hohokam ballcourts (large, open structures of apparent ceremonial use) required the labor of hundreds if not thousands of individuals, and the increasing prevalence of these architectural features suggests increases in overall population. The magnitude of population necessary to accomplish these tasks certainly must have been in the thousands, perhaps even tens of thousands, a range that, with further examination, proves almost certain.[23]

The third method of estimating population considers the numbers of individuals who can be supported by an acre of irrigated land. This method relies on the analogy of the subsistence patterns of ancient people with historic native populations. Of course, difficulty in the analogy arises in terms of modern practices that diverge from the ancient regime. For example, domesticated grazing animals increase water demand by at least ten times over the needs of people. Cattle, horses, and domesticated sheep, on the other hand, were absent from the Hohokam world.[24] Other differences include variations in agricultural practices, such as evidence suggesting that the Hohokam cultivated agave and other native plants, while the O'odham simply gather them. And last, there are cultigens of the historic period which have no corollary in the prehistoric—the primary example being wheat. For example, wheat typically requires 1.77 acre-feet of water per acre, while corn requires 2.76 acre-feet of water per acre. Thus wheat is somewhat less demanding of scarce water resources. In general, however, water demand increased with the introduction of the wide range of European cultigens, such as "wheat, barley, oats, citrus fruits, apples, apricots, pears, grapes, chickpeas, carrots, radishes, and onions."[25]

Some researchers have suggested that a combined approach of examining living sites in conjunction with irrigated acreage might provide some degree of reliability in population estimates. At least such a combination of factors can offer some hard numbers for our contemplation. Paul and Suzanne Fish followed such a line of reasoning in their study of the Phoenix Basin in central Arizona. Similar reasoning can be used to determine the population in the Tucson Basin.

First in the Fishes' equation is the spacing of platform mounds along the Salt River. Assuming that "communities integrate all adjacent territory, whether irrigated or not," the twenty-three mounds in the Phoenix Basin each occupied over fourteen square miles. The irrigated area within the fourteen square miles was over 2,300 acres. In comparison, the mounds in the Casa Grande system along the Gila River each irrigated a comparable 2,500 acres, and were similarly spaced along the river.[26]

With this fairly certain understanding of the area controlled by the communities, population numbers start to tease. All that is required is faith in the viability of the analogy between historic and prehistoric cultures, and an awareness of the basic problem in the use of living sites to determine population numbers; namely, how many of the platform mounds were being occupied simultaneously? Observations of the Pima Indians indicate that families comprising five members subsisted on 2.12 to 5.31 acres of irrigated land. Using these numbers, the math is easy. If each platform mound irrigated 2,471 acres (an even 1,000 hectares), then a population results in a range from 2,325 to 5,825 persons. Using these estimates for populations of individual mound communities, a total for the twenty-three mound communities in the Phoenix area ranges from 52,900 to 133,400 individuals.[27] Did the twenty-three mound communities all exist simultaneously? Even if only half were inhabited at any one time, the population levels still range from 25,000 to 65,000 persons. If one-quarter were inhabited (five or six mounds), then the range falls to 12,500 to 32,500. This estimate, arbitrarily divided by four, falls roughly within the "reluctant" estimate of 15,000 to 30,000 offered by the participants in a "regional systems" seminar on Chaco and Hohokam cultures. David Doyel estimated the population of the Phoenix Basin during the middle Classic Period to be 24,000.[28]

Even these conservative estimates result in populations in the tens of thousands. Moreover, underlying assumptions about the Hohokam's efficient utilization of their environment should push the estimates upward, especially because the Salt River's flow *could* have supported the higher numbers. In fact, the Salt River could have supported all twenty-three platform mounds simultaneously (of course, had the Hohokam exploited the river system to its fullest, they would have left themselves very vulnerable to fluctuations or diminutions in the river's flow).

Ultimately, political agendas may shape population estimates as much as scientific analysis. A recent study of Arizona's rivers cites the highest estimate, and actually inflates the numbers slightly—50,000 to 200,000—perhaps as a way of emphasizing the decline of the river from earlier, valorized times, whether or not the numbers coincide with the most-respected archaeological thinking.[29]

The Tucson Basin, through which the Santa Cruz River flowed, supported a "peripheral" segment of the Hohokam population. Nonetheless, a significant population inhabited the area. Experts have categorized many of the differences and similarities between the core and periphery areas, such as the spacing between villages. In the Santa Cruz River valley villages were spaced closer together than in the core areas (about every 1.8 miles in the Tucson Basin versus 3.3 miles in the Phoenix Basin), but since the Santa Cruz River villages probably incorporated more of the "lateral hinterland" (terrain beyond the floodplain, such as terraces and bajada slopes), population densities were similar.[30] The number and type of villages along the forty-eight miles of river in the Tucson Basin varied during the Classic Period as the population moved from a dispersed pattern of smaller villages to a more dense pattern of platform mounds and larger villages. During the Tucson phase of the Classic Period, the Hohokam occupied seven platform mounds, three large villages, and ten smaller villages.[31] Given these densities and relying on the same methods described above for the Phoenix Basin, the population in the Tucson Basin ranged between 20,100 and 44,600 individuals during the Classic Period (or, to appear more sensible, this range estimate divided by four equals 5,025 to 10,150). This cautious range conforms to a recent population estimate shared by William Doelle, the foremost authority on the Hohokam in the Tucson Basin: "By roughly 1,000 B.C. the possible ancestors of the Hohokam had constructed an irrigation system along the Santa Cruz River. Wells that tapped shallow ground water were dug at least as long ago as 400 to 800 B.C. Population size for these Early Agricultural Period groups is not yet known, however, it has been estimated that a peak population of 6,000 to 7,000 was probably reached around A.D. 1000."[32]

To place these numbers in context, Tucson's population in 1864 was 1,526, far smaller than the lowest Hohokam estimate, and a smaller population even than the beleaguered Spanish maintained in 1800. In 1890, Tucson's population of about 5,000 equaled the lowest, most conservative estimate for the Hohokam population. The Tucson Basin did not surpass a population of 10,000 until the 1910 census, when it reached 13,193. The modern city reached the highest range of Hohokam population in the 1920s and 1930s. The U.S. census counted about 20,000 residents in Pima County in 1920 and 32,000 residents in 1930.[33]

How close does the high range from about 20,000 to 45,000 come to the maximum sustainable population in the Tucson Basin? As previously stated, the maximum Hohokam population seems to have coincided with increased precipitation around A.D. 1000 and the spread of dry farming onto the terraces and bajada slopes near the river. Was this the practice that stretched the Hohokam population beyond the ability of the ecosystem to support it? Does the story of Hohokam decline serve as the poster child for declensionist environmental narratives? Possible scenarios explaining the Hohokam disappearance include warfare in the 1300s, or flooding—huge floods occurred on the Salt River in the 1300s—which may have irreparably destroyed the Hohokam canal system. In the Tucson Basin a plausible explanation points to the arroyoization of the river, which may account for the Hohokam's abandonment of the area. The entrenchment of the river through most of the basin could have disrupted the central pattern of Hohokam agriculture. As the river channel deepened, the flow of water fell below the intake gates to the Indian's fields, thus leaving the agricultural areas high and dry. According to this scenario, the paleo-arroyo gouged a channel through the floodplain, and it was not until the arroyo filled in (aggraded) during the fifteenth or sixteenth centuries that the pre-contact farmers returned to the river valley.[34]

A constant in the human history of the bioregion is the river. Over geologic time, the river has changed in size, configuration, and perhaps even direction. But through the millennia of Early Agricultural and Hohokam presence, the river apparently remained a fairly constant and reliable source of subsistence, and the native society grew steadily over the centuries to levels unmatched until the twentieth century. If, as Julio Betancourt proposes, the river became entrenched at that time of Hohokam maximum population, the amount of water in the river could have still supported the population, but its availability for intensive agriculture would have decreased to the point of disaster for the culture.

Within a few hundred years the river had returned nearly to its former condition, having migrated a mile or two to the west (perhaps as the result of Hohokam engineering), but meandering again within a shallow course suitable for floodplain farming. And appropriately enough, human inhabitants returned to the basin. It is even safe to assume that the Pima Indians who now inhabited the basin, akin to the earlier Hohokam people, began a gradual process of steadily increasing populations. But before the Pima population could grow to anywhere near the Hohokam levels, the Spanish arrived with their own particular designs and requirements for the river and the valley's water resources.

The Hohokam perception seems certainly archaic and animistic, but their manipulation of the river suggests an approach to a modern outlook. The digging of wells and extensive irrigation canals sets the Hohokam culture apart from other prehistoric and even historic inhabitants. Possibly changing the course of the river furthers the Hohokam's approach to a modernistic character. From the moment of Hohokam decline, however, during the intervening centuries before the presence of an industrialized society on the river, climate and geology once again came to dominate the river's history, with only occasional and tangential human influences affecting the river.

3

Cattle, Wheat, and Peace

1500–1820

A GAP OF 200 YEARS OCCURRED between the collapse of the Hohokam culture between A.D. 1450 and 1500 and the arrival of Spanish settlers in southern Arizona toward the end of the seventeenth century. The first Spanish explorers in the region traveled in close proximity to the Santa Cruz River in 1540 when Coronado's expedition passed through southern Arizona in search of the golden cities of Cíbola. In their passage from the south, Coronado's party may have traveled up the San Pedro River, which flows from south to north in the valley just east of the Santa Cruz River valley. From the San Pedro, the Spaniards traveled north and east toward the Rio Grande River valley, which eventually became the center of Spanish settlement in the region. No permanent colonization or missionary effort appeared in the Santa Cruz Valley for another 150 years. By the time the first Spanish settlers finally arrived in the 1680s and 1690s, Santa Fe, then the hub of Spanish settlement in New Mexico, had been in existence for about eighty years.

In the Santa Cruz Valley at the turn of the eighteenth century, the largest native population lived in the Tucson Basin, with other sedentary Indian villages interspersed through the lower and upper basins of the valley. The Spanish established their missions and *visitas* (places visited by Spanish priests on a semiregular basis) at the sites of these villages, mostly in the upper basin, at Tubac, Tumacácori, Calabazas (near the confluence of Sonoita Creek and the Santa Cruz River), Guevavi (near present-day Nogales), and at Santa María Soamca (near the present site of Santa Cruz, Sonora).[1] No permanent Indian settlements remained on the Salt River, and only a few small settlements persisted on the Gila—in other words, the Phoenix Basin core Hohokam culture had totally collapsed.[2]

Did the Spanish bring a new perception of the river, differing markedly from the Pima or Hohokam vision of the valley's water resources? The Spanish did bring with them new cultigens and grazing livestock, as well as an aspect of commercialism, all of which affected the river by placing added demands on the surface flow. However, the perception of the valley's water resources held by the Spanish settlers was consistent with the archaic view held by the Pima Indians residing in the valley. Even though the Spanish culture differed greatly from that of the indigenous Indians, their attitude toward the river—their perception of its extent and nature—matched the Indians' view of the river's scarcity, precociousness, and centrality to life. The Spanish settlers, familiar with semiarid conditions in their homeland, sought to reside within the natural parameters determined by the desert ecosystem, intent in the meantime on wringing every drop of useable water from the surface flow of the stream (albeit with much less success than the Hohokam). The river continued its pattern of ebb and flow during the Spanish period, subjected to new pressures from livestock grazing and intensive agriculture, but the perception of the new humans inhabiting the valley resembled that of the Hohokam and Pima in its concern with surface flows as they naturally occurred.

Father Eusebio Francisco Kino established several missions in the river valley, and he and his companion, Juan Mateo Manje,[3] recorded the numbers and circumstances of the native inhabitants. Although problematic in their particulars, these Spanish censuses give an indication of the native population according to the Europeans. It should be noted that there was a general tendency among Spanish officials, both priests and military leaders, either to inflate estimates of local conditions or exaggerate problems—whichever they expected would be most effective—when angling for support from state or national governors and institutions. Father Kino, for example, was said to possess a considerable ego, and contemporary critics accused him of inflating reports of his own success in converting the natives and developing the region. Historian Herbert Bolton put a positive spin on the controversy: "The clamors of the faint-hearted, however, merely served to bring out that optimism which was one of Kino's strongest qualities, and in his letters to the viceroy he discounts the dismal prophecies of the malcontents." John Kessell described Kino's "optimism" in slightly different terms: "self-effacement was not a distinguishing characteristic of the Apostle to the Pimas."[4]

Manje estimated that 2,170 Indians lived in the entire river valley in 1697. The year before, Kino had reported 830 "Papagos" ("bean eaters" or "bean people") at the village of Bac living in 176 houses (later the site of the mission San Xavier del Bac) and 758 Indians occupying 177 houses at the

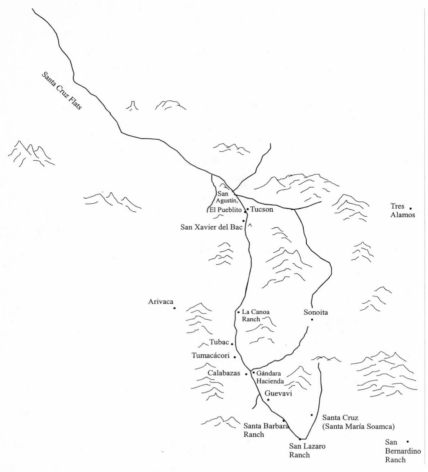

Spanish missionaries such as Father Eusebio Francisco Kino generally established missions and *visitas* at the site of existing Indian villages. Such was the case in the Santa Cruz River valley, where the missions, visitas, and later *presidios* mirrored Indian settlement patterns. Tucson became the northern outpost on this section of the Spanish frontier established by Irish expatriate and Spanish commander Hugo O'Conor in 1775.

village of San Agustín (on the west bank of the Santa Cruz River seven miles downstream from Bac, later the site of the Spanish presidio of Tucson on the river's east bank). Thus the total population in the middle basin of the river valley was reported by Kino to be 1,588. In 1699 Spanish officials counted 900 "Pimas" at San Xavier and another 800 Indians at San Agustín, a slight increase from 1,588 to 1,700, although the round numbers may cause some question as to their accuracy. Archaeologists Czaplicki and Mayberry maintain that, overall, the historical record suggests an Indian population in the Tucson Basin in the seventeenth century of about 2,580 persons, admittedly a population mostly missing from the archaeological record.[5]

The Indian population in the upper basin was also significant and also generally missing from the archaeological record. Archaeologists have studied two sites near Tumacácori that were inhabited beginning in 2000 B.C., but many if not most of the local natives at the time of Spanish contact inhabited *rancherías* (transitory villages situated for seasonal agriculture and other subsistence activities) that were scattered throughout the area and that left no trace for future archaeologists once the inhabitants had died or moved on. While there were about 200 Indians at Guevavi in 1700, there were hundreds more in rancherías and smaller villages scattered throughout the vicinity. By 1732 the Indian population at Guevavi and its four visitas was "something over 1,400 souls."[6]

For the entire region of the Pimería Alta on the northern frontier, Kino had estimated a total of "10,000 souls." Kessell estimated about 25,000 natives spread across the Pimería Alta's 50,000 square miles. This was obviously a very sizeable population, although dispersed over a wide area, of which the Santa Cruz River valley was only a part. The Hohokam, on the other hand, had reached their population maximums in a more concentrated and dense living pattern along the banks of the area's rivers.[7]

The impact of Spanish settlement on the bioregion was limited through its first fifty years, primarily due to the limited Spanish presence and pattern of settlement, which mirrored the native villages and their subsistence practices. This is not to say the Spanish "conquest" in the Santa Cruz River valley was less disastrous for the native people than in other regions. The familiar story of exploitation, disease, and cultural degradation took place in the Santa Cruz Valley as well. Over time, Spanish settlement resulted in a wide range of environmental effects, including deforestation and overgrazing, introduction of new plant cultigens and their respective insect communities, wind erosion and desertification, and the ever-present manifestations of disease, but most of these impacts remained slight until the mid-eighteenth century.[8] Environmental effects of Spanish colonization occurred earlier in the Rio Grande Valley of New Mexico and in Texas simply because the Spanish settled there earlier. Scholars such as Robert MacCameron and Jesus de la Teja have studied the Spanish in New Mexico and Texas and have found similarities in the environmental effects in both places, but the specific histories vary due to differences in terrain and climate. Likewise, the history of Spanish colonization in the Santa Cruz

River valley follows its own trajectory within the general parameters of Spanish colonization.[9]

As to the Spaniards' perception of the river, a vision of its meager surface flow dominates their reports and descriptions. Kino remarked that the river's flow could support a city of 5,000, a number his critics challenged as greatly inflated. To place this estimate in context, the population of Santa Fe in 1680 prior to the Pueblo Revolt was 2,900. The population in the upper Rio Grande Valley did not reach 5,000 until the mid-1700s. Colonial Boston's population in 1690 was about 6,000, and about 13,000 in 1770. Although the Spanish dug wells and constructed irrigation canals (which are discussed below), from all indications the Spanish efforts at exploiting the river went no further than the surface flow would allow.[10]

Although aware of the limits of their environment, did the Spanish attempt nonetheless to dominate nature in their new home? For the pre-industrial Spanish, no domination of nature was possible, even if they had attempted or desired it. The Hohokam had exerted a tremendous influence on the river's circumstance without any apparent dualistic concept of human society separate from nature, but from the river's perspective, indigenous animism or European Christianity made little difference, thus my use of the "archaic" label for the pre-industrial societies includes the Spanish. Whether they hoped to dominate nature or not, they failed to do so, possessing neither the engineering skill nor the technological tools to accomplish such a goal.[11]

Kino first entered Arizona in 1691 when he visited the Indian village at Tumacácori, but he may have been preceded in the Santa Cruz River valley by a Spanish colonist, José Romo de Vivar, who established cattle ranching in the region in the 1680s.[12] More Spanish colonists moved northward into the Pimería Alta through the early 1700s, so that by the 1730s church records indicate a burgeoning Spanish presence. James Officer summarized Kessell and other sources: "Although it has not been established that there were Spanish settlers in what is now southern Arizona in 1732, there certainly were some nearby and they were not long in going northward. In testimony that he gave following the Pima Revolt of 1751, don Nicolás Romero stated that he had settled in the San Luis valley twenty-five miles south of Guevavi in the 1720s. . . . Before the 1730s ended, the written records of the Guevavi mission had come to include regular references to Spanish families of the region."[13] In the first fifty years of colonization, the Spanish had begun cattle ranching and mining and had introduced the cultivation of wheat, but the level of ranching and farming activity had been severely limited by Apache raids. Nonetheless, evidence of the changing agricultural patterns that would later impact the river's flow was clearly evident by the 1760s.

Father Kino and the Spanish settlers both conformed to and altered traditional methods of agriculture in the semiarid environment. Kino was first of all a Jesuit missionary, but he was also a determined rancher. The Spanish drove cattle and sheep north from ranches in southern Sonora to stock the missions and presidios of the Santa Cruz River valley. Kino stocked a ranch near the mission at Caborca with over 100 head of cattle and 100 head of sheep, and in 1700 he brought 1,400 head of cattle to Bac. Historians also give him credit for the introduction of wheat cultivation in the area. Part of his mission to the natives was to encourage more intensive cultivation of the river valleys, with an emphasis on wheat but never to the exclusion of corn.[14] During a visit to the Indian villages in the middle basin in 1698, Kino reported on the Indians' harvest of "50 *fanegas* [a Spanish bushel, or fanega, is the equivalent of 3.2 English bushels][15] of wheat and corn. They were undertaking a new planting of wheat, all for the Father who for years they have been wishing to receive. . . . [The cattle] were very fat, the pasturage being very abundant." At San Agustín del Oyaur [in the vicinity of future Tucson] Kino reported many more fields planted in corn, squash, beans, and watermelons. And in 1700 when the mission was established at San Xavier del Bac, the Indians opened new agricultural fields that in April caused Kino to exult in his journal, "a good field of wheat which was beginning to head." Officer also reported the truism held by the priests, "Indians do not come to Christian service when they do not see the maize pot boiling." On the other hand, Meyer reported that grazing cattle sometimes thwarted the missionaries' work: "The Papagos at Father Kino's Mission of Remedios resisted Christianization because the missionaries pastured so many cattle that the watering places were drying up." To accomplish the intensive cultivation of grain and vegetables, the Spanish had built crude irrigation canals, as is clear from various references to and drawings of Spanish *acequias* along the Santa Cruz River. In no measure, however, did the Spanish systems come close to matching the sophistication or effectiveness of the Hohokam canals. Meyer characterized the Spanish acequias as occasionally well-built but most often crude and inefficient. Generally smaller and poorly designed, the canals at times "served no function at all."[16]

The Spanish basically adopted the Indian subsistence diet, supplemented by wheat, barley, oats, and European varieties of vegetables and fruit, crops that had the aggregate effect of increasing water demand. Father Ignacio Pfefferkorn gave the most-detailed accounting of the Spanish and Indian diets, which included the quite adaptable pinole. Indians prepared pinole from mesquite beans and corn, and with the arrival of the Spanish, from wheat as well. Modern-day connoisseurs of Mexican food can attest to the variety of dishes ensuing from Indian and Spanish culture derived from both corn and wheat. Pfefferkorn also mentioned the settlers' taste for meat. The Indians readily consumed horse and mule meat but had "a loathing for pork." Beef was the usual fare, with chicken and mutton available for the wealthy. And the inevitable sweet tooth existed in the desert. The Spanish craved chocolate, but its expense made it rare. The Indians had a sweet tooth too. At one point the Tucson commander sought to pacify a group of peace-inclined Apaches with a supply of "raw sugar." Alcohol was available on the frontier also, primarily in the indigenous forms consumed by Indians for centuries: mescal, tesquino, and Papago wine. Sonora had its own variety of mescal, *bacanora,* which the local commander taxed in the Tucson presidio, perhaps indicating its prevalence.[17]

The most-efficient farmers during the Spanish and Mexican period were always the sedentary, "village-dwelling" Indians. The Spanish brought with them a wide variety of crops from Europe, Mexico, and Central America—everything from citrus to chickpeas, radishes, carrots, and onions—but not all these crops were grown in every location at all times. For example, the Indians in the Santa Cruz Valley took to the cultivation of wheat very quickly, but fruit trees remained limited to the priests' garden at San Xavier del Bac for many years.[18]

By all accounts, the Spanish seemed to be more interested in stock raising. Father Pfefferkorn, of German descent, expressed a clear disdain for his Spanish charges, commenting that "not even the lowliest can be expected to make a journey on foot, be the trip ever so short," and that most settlers "farm in such a lackadaisical manner that a foreigner cannot observe them without becoming indignant. That they derive their substinance [*sic*] from farming is due not to their labor but to the exceptional fertility of the soil." The Jesuit priest could not deny their interest in stock raising, however: "The only occupation which is to their liking is the raising of livestock. In that work they are really indefatigable."[19]

If the Spanish were half-hearted farmers, another group in the region had grown increasingly disdainful of sedentary agriculture or ranching and many other forms of pastoralism. The Apaches had once practiced a pattern of seasonal and migratory agriculture, but incessant warfare with the Comanches had turned the Apaches into a nomadic culture of marauding bands, deriving much of their subsistence by raiding the stores of sedentary Indians in the valley and the colonizing Spanish. The Apaches had entered the region from the north and east during the 1700s, spreading into southern Arizona from New Mexico. In the warfare that ensued, the sedentary Indians often allied with the Spanish against the Apaches.[20]

As expected, given the fertility of the river valley and the farming skills of the sedentary Indians, and not withstanding the losses to Apache raids, agricultural surpluses occurred, and the Spanish put these to use in support of new building projects. For example, when the Spanish sought to organize the Indians in 1768 to construct a breastwork at Tucson for defense against the Apaches, they had to promise wheat to construction workers and their families, since the workers were being pulled away from subsistence activities. During the fourteen years of construction on the church at San Xavier del Bac before its completion in 1797, laborers and craftsmen had to be supported. Even after the craftsmen completed the church, the supervising priests assigned the workforce to new building projects near the presidio. In a more prosaic sense, the increasing herds of livestock required surplus grain beyond that needed for human consumption, or in the case of the church at San Xavier, for spiritual and artistic expression. The common complaint from the Indians became a virtual litany as wandering Spanish livestock foraged among the Indians' fields.[21]

Through the early 1700s livestock populations, although troublesome for Spanish-Indian relations, did not seriously interrupt the flow of the river. The Santa Cruz could support large human populations or large livestock populations, but not both. During the first fifty years of Spanish presence, from roughly 1690 to 1740, human populations were small enough to allow livestock to graze without overtaxing the river's flow. In addition, raiding Apaches repeatedly winnowed the herds of cattle and horses, keeping livestock populations in check. However, the next twenty-five years witnessed the transition to environmental stress. Through the middle decades of the eighteenth century, the Spanish settlements in the Santa Cruz River valley lost the balance between human and livestock populations. The Apache raids continued, but they mainly had the effect of concentrating the human population in the valley around the military presidios. With the concentrated population in the valley, livestock populations also increased. The Apaches still raided the herds, but the purloined livestock often never left the river valley. Even when thinned, livestock herds placed increasing demands on water resources, and the springs where cattle, horses, and sheep watered began drying up. By the 1760s, below-average rainfall coupled with continuing Apache raids created a crisis situation for the Spanish.[22]

One result of the crisis was the Spanish policy in 1761 to relocate the Sobaipuri Indians from the San Pedro Valley to the Santa Cruz Valley. The Sobaipuri Indians had become increasingly vulnerable to Apache raids, and their consolidation with the population in the Santa Cruz Valley would simplify the Spanish defense. So whereas the total population in the region declined due to Apache raids, in the Santa Cruz Valley the population was increasing, placing added demands on the Santa Cruz River while relieving some of the demand on the San Pedro River. The Spanish placed most of the Sobaipuri—the number is unclear, perhaps 200 or 400 individuals—at Tucson, with others established at Santa María Soamca and Sonoita. In 1763 the Tubac commander, Captain Juan Bautista de Anza, furthered the policy of retrenchment by approving the abandonment of several ranches in northern Sonora. Most of the settlers moved from the isolated ranches to the protection of the presidios.[23]

A drought in the 1760s added to the difficulties faced by the valley's residents. The Spanish livestock began to deplete water supplies, so that Indians downstream from the mission herds complained that their water supply was dwindling. In the particularly dry year of 1761 the priest at San Xavier del Bac, Father Manuel de Aguirre, wrote to the governor that there was "plenty of land for everyone, but not enough water to sustain the existing Spanish and Indian population." This was the first recorded incidence of insufficient surface flow in the river to satisfy demands placed upon it by human society. In times of normal rainfall the river still met all the demands placed upon it, but in the inevitable times of below-average precipitation the river's flow was inadequate. The timing of rain was also critical. The summer rains arrived "very late" in 1768 in the Tucson Basin, disrupting native agriculture to the point that the Indians became more amenable to proselytizing by the Spanish priests.[24]

At the time of Jesuit expulsion in 1767 (a result of the political struggle between King Carlos IV and the Pope), and the arrival of their Franciscan replacements, most Hispanic colonists in the region had converged on Tubac in the upper basin. John Kessell described the presidio's population as "close to five hundred." The garrison was composed of "fifty-one men, including three officers and the interim chaplain, their dependents, servants, and assorted hangers-on." The civilian population included "dozens of settlers . . . many of them refugees from the abandoned ranchos upriver." The census of civilians counted "34 heads of family, 144 dependents, plus 26 servants and their families, for a total of well over two hundred." James Officer described the population at Tubac at this time as probably "the largest it would ever be during the Spanish and Mexican periods." Perhaps facilitating these larger population levels, a change in agricultural practices accompanied the replacement of the Jesuits by the Franciscans. Productivity in agriculture increased as some of the communal mission lands were turned over to the Indians as individual plots (*milpas*).[25]

The overall population throughout the region continued to decline through the 1770s as Apache raids increased and dry-cycle conditions persisted. The shrinking population continued to concentrate around Tubac for their protection. By 1773 the Franciscans had abandoned the church at Guevavi, built by the Jesuits prior to their expulsion in 1767. In the next year the Indians residing at Guevavi and at the visita at Sonoita joined the movement to the relative safety of Tubac and its environs. After these migrations, more than 200 Indians resided at Tumacácori and Calabazas (98 and 138 respectively), adding to the burgeoning population in the vicinity of Tubac (at this time 399 Indians and no Spanish were living at San Xavier). In the face of the continuing drought in 1774, Tubac commander de Anza instituted "a weekly water rotation" between the mission and Indians at Tumacácori and the presidio downstream at Tubac.[26] Spanish officials earlier had debated the wisdom of combining the Indian village at Calabazas and the mission at Tumacácori given the shortage of irrigated land. But a shortage of water was not at fault, at least according to one official. Rather it was a dearth of "industry" caused by the threat of Apache raids: "neither at the mission nor at the visita do they have enough land under irrigation for their combined support. Even though it is true that at Mission Tumacácori with more industry they could provide sufficient irrigation for the support of both pueblos, the constant hostility of the Apaches prevents it."[27]

In the meantime, the Spanish military had encouraged the Indians at Tucson to build an "earthen breastwork" for their defense, which was the first step in the establishment of the presidio at Tucson in 1775. In that year Spanish authorities ordered the transfer of the garrison from Tubac to the new presidio to the north, and Tucson became the center of Spanish defense of the Pimería Alta.[28]

The shift of the Tubac garrison to Tucson, including many of the civilian inhabitants, shifted demands on the river from the upper basin to the middle basin.[29] The Tucson commander, Don Pedro de Allande y Saabedra, began making land grants to some of the civilian arrivals, and this created competition with the Pima village across the river for the dwindling water supply.[30] Also, as the Spanish presence in Tucson increased, livestock herds expanded, and soon the endemic Spanish-Indian dispute arose as cattle began trampling the Indians' crops.

The demand on the river through the colonial period followed no simple, or linear, pattern. It increased and slackened, varying from basin to basin, and changed as a result of the varying success of Spanish and Mexican settlement. The climate and weather patterns remained ever-present variables in the river's changing circumstances, but now the pattern of Spanish settlement, with all of its tortured ups and downs, began to influence the river at different times and in different places on a scale to rival even the weather.

The advent of declining water supplies and increasing environmental stress in the river valley seems clearly evident by 1761, when San Xavier's priest complained of insufficient water. But the pattern was not a constant or steady decline. Periods of bountiful rain, or decreasing demand on the river, might bring it back to flowing abundance. Thus the accounts by Spanish observers sometimes baffle in their inconsistency. One example is the reference to the terrain around Tucson during the May 1782 attack on the presidio by the Apaches. By 1782 the dry cycle of the 1770s had ended and the river had returned to flowing abundance. The reports of the battle indicate that the Santa Cruz at that time in the Tucson Basin was divided into two streams which bounded to the east and west of the Indian village of El Pueblito west of the presidio. The main channel of the river was apparently the eastern branch, between the village and the presidio. A permanent bridge spanned the stream at that point, providing the primary access to the presidio from the Indian town. During the battle in 1782 presidio troops had defended the bridge, and the Apache assault was stymied there. Of interest is the fact that the Apache warriors, many on horseback, made no attempt to outflank the Spanish defenders by crossing the river upstream or downstream, apparently due to the depth and width of the channel. Ethnohistorian Henry Dobyns recognized the significance of Spanish accounts of the battle as a comment on the status of the river: "The fact that this heavy streamflow existed on the first day of May emphasizes the difference between the aquatic ecology of the Santa Cruz River valley in 1782 and that of a century and a half later. The month of May is part of the driest season of the year in southern Arizona, after the winter rains have ended and long before summer thunder showers begin. While the hydrology of the area must be inferred from brief references to the streamflow in accounts of the battle, these references are quite clear."[31]

Dobyns understood the significance of the accounts, but he was mistaken in his assumption of a linear pattern of decline for the river from the 1780s onward. The earlier accounts of stress on the river make clear that the river varied in its circumstances as the forces acting on it exercised differing influences. From insufficient flow in the 1760s, to impressive currents in the 1780s, decline had been followed by surplus. Conversely, the apparent abundance of water in the 1780s turned to scarcity in the early 1800s, then flowed again to sufficiency in the 1830s and 1840s. Ebb and flow, decline and rise, no linear pattern emerges in the history of the river's circumstances. Increasing agriculture and proliferating livestock served as powerful influences on the river. Increasing livestock herds also brought the added potential of overgrazing the watershed, trampling the banks, and creating pollution concerns. But agriculture and ranching activity in the valley ebbed and flowed, generally in relation to increases or decreases in Apache raids, and the effects of settlement varied over time in the valley.[32]

The Spanish perception of the river remained consistent with native residents, but water politics became visible during the Spanish period, unlike the hidden politics of native, prehistoric societies. It seems safe to assume that the Hohokam and Pima residents of the valley engaged in political discussions regarding use of the river. Since these native societies required cooperative efforts to achieve subsistence, some sort of political consensus must have developed. With the Spanish presence, political decisions affecting the river became explicit.

The "New Indian Policy," or *establecimientos de paz,* instituted by the colonial government in the 1780s proved fateful for the history of the river. The inability of the Spanish colonial government to control the Apaches meant a setback for settlement in the region. With the new policy, Apache raids declined and a new level of development took place. The increases in settlement in the valley renewed the pressure on the river, and officials soon began describing the river in terms of insufficient surface flow.

The New Indian Policy came about due to the Spanish viceroy's mounting frustration with expensive and ineffective military efforts against the hostile Indians. Despite a constant military presence, Apache raids on

Spanish settlements continued into the 1780s. In one typical raid, the Apaches attacked the presidio at Tucson in 1784, killing five soldiers and escaping with 150 horses and 40 mules. While Spanish commanders reported success in campaigns against the Apaches, the raids continued nonetheless. Finally in 1786, Viceroy Bernardo Gálvez announced the new policy, which turned out to be more effective in quelling hostile activity than any previous military campaign or program. Quite simply, the New Indian Policy "bought off" the Apaches, paying the Indians to stop raiding by providing them with the goods and articles they desired in the first place: "horses, guns, food and ornaments." In addition, the New Indian Policy sought to foment internal divisions within the Apaches by dispensing firearms and alcohol to groups in such a manner as to create their economic dependence on the Spanish. The New Indian Policy also advocated concerted military operations against hostile Indians who did not agree to the pay-off terms. This eventually resulted in the taking of many Apache captives, mostly women and children, some of whom were enslaved by their Indian and Spanish captors.[33]

The new policy got off to a shaky start, however, when a group of Apaches took advantage of the policy's initiatives to infiltrate and then raid the Pima community at Calabazas in October 1786. The soldiers from Tucson eventually tracked down the Apaches and defeated them in a quick skirmish, but in the meantime the Pimas at Calabazas abandoned the village. Eventually the New Indian Policy turned the tide against the Apaches, and raids began subsiding. In 1787 the Spanish reestablished the presidio at Tubac, and within a few years Apaches seeking peace began showing up at Tubac and Tucson. As mentioned previously, in Tucson the commander welcomed one group of Apaches with a gift of "raw sugar," and bought fifty head of cattle "to provide meat for the new arrivals." Quickly the number of Apaches residing outside the presidio walls grew to over one hundred. The Apaches lived on rations provided by the presidio and probably never engaged in serious farming, in part due to their disdain for pastoralism, and in part due to the new policy's intent to keep them dependent on Spanish largess. The presence of an increasing, but nonproductive, population in the middle basin added to the necessity for surplus production for sale to the presidio. This proto-commercialism eventually impacted the type and level of agriculture in the basin, and the subsequent stress on the river.[34]

The addition of a new Indian population to the civilian and military population in the Tucson area in the 1770s created serious pressure on the river. As Michael Meyer described, there were four competing entities utilizing the river's water in the Tucson Basin during the late colonial period: the Pima village known as El Pueblito near the presidio, the presidio community, the "communal mission lands" of San Xavier del Bac, and the individual farms of the Pimas and Papagos. The conflict between these constituencies eventually resulted in "an accord." The priests in the area were sympathetic to the Indians' claims, and as a result, the settlement dictated by the Sonoran government was quite favorable to the Indians. Spanish officials allotted the Indians "three fourths and the Spanish one quarter of the waters from an important spring upon which all agriculture in the immediate area was dependent."[35] Given the dwindling resource in the face of increasing demand, however, the settlement simply papered over a growing controversy.

As a result of the New Indian Policy and the reestablishment of a presidio at Tubac, and despite renewed drought conditions, human and livestock populations in the valley increased through the 1790s. The relatively peaceful relations between the Apaches and the Spanish and sedentary Indians relieved a major impediment to growing populations and increasing settlement. When Franciscan Friar Bringas traveled through the region in 1795 recording the circumstances he found in the Pimería Alta, his report oftentimes dealt with the increasing conflict between colonists and Indians over water and arable land. Bringas reported that settlers blatantly disregarded the earlier agreement allotting the Indians three-fourths of the water. Ultimately, the royal authorities ignored Bringas and his report. An incident occurred in Tucson just after the Bringas report that illustrates the increasingly intractable situation. In January 1796, drought forced a group of 134 Papagos from northwest of the presidio at Tucson to resettle near the Pima village of El Pueblito at the foot of Sentinel Peak (or perhaps San Xavier).[36] The increased population and drought conditions meant there was no longer sufficient water in the river to satisfy all needs. The Spanish would not share the limited water with the new arrivals, and to add hardship to injury, the Papagos repeated the common complaint that Spanish livestock was damaging their meager crops. Later, as drought conditions intensified, presidio soldiers simply helped themselves to the water, "leaving the Indians hardly a trickle."[37] The conflict between the Indian and Spanish over the water allocation continued into the Mexican period. Resolution of the dispute proved elusive, to say the least.

Nascent commerce was on the rise as the nineteenth century arrived, and this proved to be another factor in the river's stress. The modest surplus of earlier decades became pronounced, such that by 1803 there were so many cattle at Tubac that the price of livestock dropped to unprofitable levels.[38] In Tucson four individuals were running pack trains bringing trade goods into the presidio from Sonoran trade centers to the south. Others were making soap, and in Tubac residents had established a weaving trade that produced 600 wool blankets that sold at five pesos each, and over 1000 yards of "coarse serge" that sold at about a half-peso per yard. Such signs of prosperity appeared in the face of drought and a river stretched to its limits. In another example of discretionary income in Tucson, Tucson presidio commander José de Zúñiga reported that locals had paid out more than 2,000 pesos for tobacco, purchased through the paymaster's office and the "company store." Tucsonenses were also paying 500 pesos annually to purchase "merchandise from the orient," which officials distributed through the Sonoran capital at Arizpe.[39]

Southern trade links flourished briefly, but the valley's isolation continually hampered development of any steady commerce with other provinces or regions. For example, in 1775 Tubac commander Juan Bautista de Anza led a detachment from the presidio on the Santa Cruz River to San Francisco Bay, but no reliable trade route developed as a result of the expedition. Not until the arrival of the railroad in 1880 would the valley experience a consistent level of commercial activity sufficient to affect the river's circumstances. Nonetheless, in the early 1800s during the establecimientos de paz, a hint of such commercial activity appeared in the valley.

In 1804 the Tucson and Tubac commanders conducted surveys as had been ordered by King Carlos IV in 1802. The 1804 census by commander Zúñiga in Tucson reported 300 "Spanish" (up from 77 in 1777) and a total population in the vicinity of the presidio including San Xavier (Spanish, mixed, and native) of 1,014. In Tubac, the second ensign of the presidio, Manuel de León, reported a population of 164, including "88 soldiers and their families, eight civilian Hispanic households and twenty Indian families." The increasing prosperity (and demand on the river) was indicated by the agricultural production reported in both Tubac and Tucson. De León reported that Tubac's annual corn harvest was about 600 fanegas (worth 1,200 pesos). The wheat crop was 1,000 fanegas (worth 2,000 pesos). The presidio also supported an impressive livestock herd. There were 1,000 head of cattle (3,000 pesos), 5,000 sheep (1,875 pesos), 600 horses (4,800 pesos), 200 mules (4,000 pesos), 15 burros (90 pesos) and 300 goats (75 pesos). Tucson produced more wheat than Tubac, about 2,800 fanegas (5,600 pesos), and about the same amount of corn, 600 Spanish bushels.[40]

Zúñiga also reported that residents produce "beans and other vegetables [that] sell at four and a half pesos a bushel." The Indians in Tucson produced cotton "for their own use." The livestock herd in Tucson was equally impressive: 3,500 cattle (10,500 pesos) and 2,500 sheep (1,300 pesos). Zúñiga listed other livestock without an estimated value: 1,200 horses, 120 mules, 30 burros, and no goats or pigs.[41]

The presidio commanders' catalogue of agricultural production and livestock proliferation indicated the largest demand for water in both the upper and middle basins to occur in the pre-industrial era. Tucson's cattle herd alone devoured much ground cover and consumed many fields of grain. They also drank lots of water.[42] Given the equivalency ratio between the water consumption of cattle and humans to be ten to one, those 3,500 cattle were drinking enough water to support, theoretically, 35,000 people. This is not meant to be a claim of absolute equivalency, but the ratio is indicative of the relative demands placed on the water supply, keeping in mind that the water supply itself varied according to geologic and climatological circumstances. If the analogy holds even partially true, then the estimate for the maximum human population supportable by the Santa Cruz River falls within the highest estimated range for the Hohokam population (20,000 to 45,000), assuming the absence of grazing domesticated animals. And, of course, the cattle had company. The sheep and horses were drinking water too.

As might be expected then, the reports by the presidio commanders describe a river still available, but much constrained. De León said the Santa Cruz flowed steadily "only in the rainy seasons. . . . During the rest of the year, it sinks into the sand in many places." De León also mentioned the Sonoita River, which flowed "steadily for the first fifteen miles of its westward course, but sinks beneath the sand seven or eight miles before joining the Santa Cruz." De León went on to describe the return of the Santa Cruz River to the surface downstream where it "collects in the marsh lands around San Xavier del Bac in great abundance." It is interesting to note that to de León, the abundant water occurred somewhere other than Tubac, despite the impressive agricultural production in Tubac supported by the fitful stream.

In Tucson, commander Zúñiga described the water resources of Tucson as tied to the rainy season, with as much accent on scarcity as bounty:

> The rivers of the region include the Santa Catalina [Rillito], five miles from the presidio, which arises from a hot spring and enjoys a

steady flow for ten miles in a northwesterly direction, but only in the rainy seasons. It is thirty three feet wide near its headwaters. Our major river, however is the Santa María Soamca [Santa Cruz], which arises ninety-five miles to the southeast from a spring near the presidio of Santa Cruz. From its origin it flows past the Santa Cruz presidio, the abandoned ranches of Divisaderos, Santa Barbara, San Luis and Buenavista, as well as the abandoned missions of Guevavi and Calabazas, the Pima mission at Tumacácori, and the Tubac presidio. When rainfall is only average or below, it flows aboveground to a point some five miles north of Tubac and goes underground all the way to San Xavier del Bac. Only during years of exceptionally heavy rainfall does it water the flat land between Tubac and San Xavier.

Zúñiga's report catalogued large agricultural production that required a constant source of water. The tone of his report, however, is pessimistic. Zúñiga noted that the land in the valley "is very fertile," but he bemoaned the poor condition of the presidio: "Why, then, do the privileged settlers not prosper more than they do?" In one example, Zúñiga noted that flourishing vineyards existed near the presidio, but the grapes were not cultivated for wine or in any other way to suit his desires: "Why do the settlers not prosper when even neglected vineyards produce a bumper crop? They hardly allow the grape to mature properly before they are selling it, to say nothing of not experimenting with new vines and cuttings."[43]

In addition to encouraging development near the presidios, the establecimientos de paz created by the New Indian Policy encouraged repopulation of areas that had been abandoned decades earlier. Spanish families had reoccupied the Calabazas area by 1808 and others returned to Arivaca in 1812. Stock grazing also returned to the San Pedro and San Rafael Valleys.[44] Hispanic families had settled these areas prior to the increasing Apache raids in the 1750s, and they returned as Apache raids declined in the 1790s and early 1800s. The ebb and flow of settlement continued through the Mexican and early American periods, with settlements expanding and contracting as the Apache wars variously raged and cooled.[45]

The economic mainstays of the Spanish in Arizona bore remarkable similarities to twentieth-century economic basics: agriculture, ranching, and mining; or in the modern vernacular, cattle, cotton, and copper. Even with these modern aspects of Spanish settlement, however, their perception of the river remained archaic, bound to the surface flow of the river. To the highest degree possible, the Spanish sought to develop the valley's prosperity, but their manipulation of the river never came near the level of intrusion—for instance, altering the course of the river—as the Hohokam possibly had attempted and achieved. The Spanish nonetheless increased the economic productivity of the river valley, and in the process increased the demand on the river's flow. In addition to renewed ranching and farming in the river valley in the early 1800s, mining resumed. The gold mine near Guevavi reopened in 1814 after a fifty-year hiatus. And likewise, miners returned to work the sites at Arivaca and El Salero, in the Santa Rita Mountains, prior to the end of the colonial period.[46]

With the increase in settlement and population, pressure rose to acquire legal title to land rather than rely on the informal arrangements between priests and settlers from the earlier period. For example, Spanish settlers at Tumacácori sought a land grant, which required that the issue of the Indians' title to the land be settled. In 1807 the Sonoran government at Arizpe issued the Indians a formal land grant of over 6,770 acres in a long strip along the Santa Cruz River originating near Tubac and extending nearly to the current Mexican border. The grant included the agricultural areas traditionally supporting the three communities upstream from Tubac: Tumacácori, Calabazas, and Guevavi.[47]

Typical of the grants obtained by the Spanish, Agustín Ortiz acquired a land grant to the Arivaca area in 1812. Arivaca was the site of an old ranch and mine, and Ortiz acquired it for 747 pesos and three *reales*. Ortiz and his family occupied the grant until at least 1833.[48]

Land grants serve as another example of political acts that had an effect on the river. By stimulating the grazing of livestock and an increase in agriculture, the grants facilitated an increased demand on the river. Another political development occurred in the early-nineteenth century—the movement for Mexican independence—and this too had an impact on the environmental circumstances of the Santa Cruz River. The movement for Mexican independence began in 1810 and marked the death knell for the establecimientos de paz. The troops on the frontier played a role in the warfare during the revolution, and the expense of fighting the war placed a drain on the Royal Treasury, which meant there were fewer troops and less money to pursue the New Indian Policy. However, nothing much seemed to happen at first; the impact of the revolutionary activity was slow in coming to Tucson and Tubac. Raids by hostile Apaches continued between 1812 and 1820, and the Spanish troops, when not engaged in political warfare, pursued the raiding Indians, but these skirmishes remained infrequent and slight compared to the earlier bloody attacks. As Officer noted, church records do not mention many deaths as the result of Apache activity in the decade prior to Mexican independence.[49] As an example of the continuing peace, in 1819 seventy-nine Pinal Apaches led by chief Chilitipage, part of a larger group of 236 Indians, settled near Tucson. Later some of the Pinal Apaches were transferred to Santa Cruz, because of "enmity" with other residents.[50]

The movement for Mexican independence came to fruition in 1820. The establecimientos de paz did not explode in a blaze of warfare, rather it slipped away through the years of turmoil leading up to Mexican independence. Revolutionary politics along with several other factors that arose during those years contributed to declining conditions on the frontier and continued to affect conditions into the Mexican period.

Epidemics returned to the river valley, as they had periodically throughout the Spanish period. New waves of disease killed more settlers and sedentary Indians than the Apaches had claimed at the height of the warfare earlier in the century. The epidemics hit the Indian populations harder than the Spanish, probably throughout the river valley, and most certainly at Tumacácori. Henry Dobyns studied the decline of the Indian population by 209 individuals at Tumacácori from 1804 to 1818. The losses to disease seem to have peaked during a smallpox epidemic in 1816 that hit all the missions in the region.[51]

As the Indian population around Tucson and the other presidios declined, arable land fell out of production, at least for a time. Spanish soldiers customarily took over these lands, planting gardens and running a few head of livestock. The soldiers and their families often occupied these lands for decades and came to consider them a possession. Some soldiers retired to the land—both a sign of the generally peaceful conditions since the 1790s (more soldiers survived to be able to retire), and the absence of the Native Americans who had previously occupied the lands. Controversy arose during the Mexican period over the title to these lands, since title often carried with it an allotment of water. As the independent Mexican government sought to administer its northern frontier, land title and water claims rose to new heights of litigation.[52]

As the establecimientos de paz receded further from view, the Apaches resumed raiding at levels unseen for thirty years. Settlements recently repopulated were abandoned once again. Land grants recently awarded fell into disuse. Cattle herds ran wild. Clearly, fewer people inhabited the river valley during the period of Spanish colonization. While the reasonable estimate by William Doelle places the peak Hohokam population of the Tucson Basin between 6,000 and 7,000, by 1800 the Spanish and native population had fallen to less than 2,000. But this was not due to the river failing to support population densities such as the Hohokam enjoyed. The river still supported bountiful life, but the forms and manner of life had changed. Cattle, sheep, and horses had replaced humans.

4

War and a Returned River

1820–1846

CONTINUITY MARKED THE PERCEPTION OF the river during the transition from Spanish to Mexican sovereignty in the valley. The Spanish and Mexican efforts to develop the region were similar, as were the ways in which they sought to put the river's surface flow to use. Although Mexican independence did not slacken the drive to develop the northern frontier, it did eventually cause the slackening of demand on the Santa Cruz River. Increasing Apache hostility combined with chaotic and fractious government policies and administration brought declining conditions of settlement in the river valley by the late 1820s. This resulted in less demand on the river and a return of the surface flow to something like its earlier meandering pattern. As always, the river in the semiarid bioregion remained dependent on variables of geology, climate, and the vagaries of human settlement.

The effort to expand agriculture and cattle grazing on the Sonoran frontier continued in the early years of Mexican independence. Facilitating these efforts was the residual effect of the establicimientos de paz, as Apache raids remained at low levels through the early years of Mexican independence. The proliferation of land grants in the early 1820s represents this continuing development process, and resulted in greatly expanded cattle grazing in the region. Settlers filed four applications for land grants on the northern frontier from 1820 to 1825, while Apache raids were still relatively limited. Two of the grants included land in the Santa Cruz River valley. The San Ignacio de la Canoa grant covered about 17,000 acres north of the Tubac presidio and extended to the middle basin near the current community of Sahuarita. The Canoa grant included the site where the river traditionally disappeared beneath the sand several miles north of Tubac (reliable water generally doubled the appraised value of land during Spanish and Mexican periods). The San Rafael de la Zanja grant included about 17,000 acres in the San Rafael Valley north of the Santa Cruz presidio. The headwaters of the Santa Cruz River flowed through this grant, watering abundant grasslands and a thriving riparian area with at least one cienega near the village of Santa Cruz. Ranching in these areas predated the land grants, but ceased in the mid-eighteenth century due to incessant Apache activity. Ranching resumed during peaceful times, and accelerated with the acquisition of land grants.[1]

The grantees began running cattle on their ranches sometimes even before they received the actual title, and grant holders often expanded grazing beyond the boundaries of their grants. Much litigation in later times concerned these "overplus" lands, as owners of former grants sought title to these traditionally utilized lands outside the boundaries of the actual grant.[2] In 1820 Ignacio Pérez applied for the largest grant in the region, San Bernardino, to the east and south of the San Pedro River. Pérez paid for the land—more than 73,000 acres—in 1822, but he never received formal title. Pérez started cattle grazing nonetheless. The stock for the San Bernardino ranch came from the mission at Tumacácori. In 1822 Pérez contracted with the priest at Tumacácori, Father Estelric, to buy 4,000 cattle at three pesos each. Of significance to the Santa Cruz River, the 4,000 cattle from Tumacácori soon began consuming water from another watershed. The transfer of the Tumacácori herd lifted a huge demand on the river's resources. From more than 5,000 head in 1820, the Tumacácori cattle herd declined by 1830 to fewer than 500 "running wild in the desert."[3]

Settlers filed five more applications for land grants in the region from 1826 to 1831. One application included the 17,000 acres of the old Buenavista Ranch, along the Santa Cruz River to the south of the Guevavi mission. These later grants occurred during a period of increasing Indian hostility, and resulted in little actual development. In 1831 the combined Sonoran and Sinaloan government enacted legislation to encourage *empresarios* to develop a large tract of land that included the Tres Alamos area on the San Pedro River. The grandiose plan primarily advocated cattle grazing, but it never got off the ground due to Apache assaults in the area. Despite the raids, Mexican ranchers continued to work the earlier grants into the 1830s. Cattle herds at Canoa and in the San Rafael Valley increased, although often in a feral fashion. Apaches targeted ranch hands (*vaqueros*), which made customary ranching practices, such as castrating male yearlings, difficult if not impossible. As a result, herds became increasingly wild in nature, dominated by increasing numbers of aggressive young bulls. This evolution of the cattle herds was complete by at least 1843, when an effort by Mexican troops to round up cattle in the San Pedro Valley failed because the cattle were too difficult to handle.[4]

Government policies and actions at the federal, state, and local levels contributed to the decline in settlement on the Mexican frontier during this time. Apache raids, which included defectors from among the peaceful Apaches near the presidios, continued in part due to the inability of Mexican authorities to continue the policies that had created the establecimientos de paz. Independence brought Mexicans freedom from the empire of Spain, but it did not bring Mexicans a stable government. The Sonoran government (for a time subsumed under the Occidente province, which combined Sonora and Sinola in 1824) never achieved fiscal solvency, and the presidio commanders complained bitterly and chronically of insufficient support. One result of the declining support for the presidios was the inability to provide payments at the former levels, primarily in the form of rations, to the peaceful Apaches. Although many of the Apaches residing near Tucson remained peaceful and loyal to the presidio commander, others left the community to join the raiding Apaches in the nearby mountains.[5]

In an attempt to compensate for the lack of funds, which left presidios perilously vulnerable, the state government enacted legislation in 1832 providing for the creation of local militia units (*los civicos*). Volunteers in the Tucson and Tubac militia later called their unit *Sección Patriótica*. Eventually these units became the primary defense against hostile Indians. Presidio soldiers during the Mexican period were rarely paid, ill-equipped, and regularly drawn away to support factions in the tumultuous state and national politics. Only infrequently were government troops able to mount significant campaigns against the Apaches.[6]

The warfare was intractable, in part due to the Mexican "bounty" policy—paying for the scalps and ears of Apache men, women and children—one of the most incompetent government policies ever devised. The Mexican government was losing the war with the Apaches, and this policy simply served to make their enemies more determined. The hatred between Apaches and Mexicans became legendary. Victories came few and far between for the Mexicans, and atrocities became commonplace on both sides. Compounding the difficulties for the Mexican government, the Apaches were not the only hostile Indians in the region. A war with the

Papagos in the 1830s culminated with a final battle near Baboquivari Peak in 1841 in which Mexican forces killed over forty Papagos and recovered a large herd of livestock.[7]

In addition to the Indian wars, Mexican officials confronted another potentially aggressive "invasion" in the mid-1820s as Anglo mountain men came down the Gila River from New Mexico, trapping beaver and seeking trade with the Pima Indians. Historian David Weber described the long recognition of the presence of beaver on the Gila and its tributaries. "In 1757 a Jesuit had even remarked that along the Gila there lived 'beavers which gnaw and throw to the ground the alder-trees and cotton woods.' "[8] The first group of Anglo trappers appeared in late 1825. James Ohio Pattie led the group, and reported with characteristic mountain-man bragga-docio to chroniclers that beaver were plentiful on the river. Pattie's group trapped thirty animals on the first night on the Gila, and harvested another 250 beaver in two weeks on the San Francisco River, a tributary of the Gila. The trappers also reported a lot of beaver on the San Pedro River, which they enthusiastically renamed "Beaver River." Three or four groups of trappers—probably ninety men in all—entered the area in the autumn and winter of 1826. The Pimas reported the presence of the group that arrived in October 1826 to the Tucson commander,[9] describing the trap-pers as "friendly," primarily due to their willingness to bestow "many gifts." These trappers may have been part of a group led by W.S. "Old Bill" Williams and Cerain St. Vrain. In December 1826, the Indians reported two groups of Americans trapping beaver on the Gila. Mexican law re-quired foreigners to report their presence to the local authorities. None of the earlier trappers did so, but three members of these later parties reported to Tucson on New Year's Eve 1826, to show the new mayor, Juan Romero, their passports.[10]

These first Anglos in the valley probably walked to Tucson from north to south up the valley from the confluence of the Santa Cruz River and the Gila, retracing the steps of the first Paleo-Indians 10,000 years earlier. Obviously these trappers viewed a much-changed scene. Drier, the river no longer reached the Gila, or even flowed much beyond the Tucson Basin, except during the most extraordinary floods. During the moist winter months, however, the trappers probably encountered seeps and springs at the base of Picacho Peak, and puddles and pools lingering after soaking winter rains. They would not have encountered a steady surface flow until reaching the middle basin of the valley, and even then would have seen no evidence of beaver. Later accounts mention beaver in the Santa Cruz River, but these reports seem fanciful. Muskrat probably inhabited the river valley, but these critters would not have inspired great awe or avarice among the trappers. No mention of the Santa Cruz River intrudes into the mountain men's accounts of the region.[11]

Trapping ended on the Gila after Indians nearly wiped out an expedition led by Michel Robidoux in January 1827. Only Robidoux, James O. Pattie, and an unnamed Frenchman survived the attack on the Salt River near modern Phoenix. Mexican officials may have breathed a sigh of relief as the foreigners lost interest in the region, but this respite from Anglo incursion was only temporary, and the difficulties for the Mexican settlements continued to mount.[12]

Another government policy that led to the declining state of Mexican settlements in the late 1820s was the expulsion of the Spanish-born priests in 1828. The *peninsulares* at San Xavier and Tumacácori had often been the only voices speaking for the native inhabitants.[13] On the other hand, in competition for scarce water, the needs of communal mission lands, to the priests, often superseded the needs of Indians farming individual or family plots. Nonetheless, the water allotment in the Tucson Basin that granted the Indians three-fourths of the water was due in large measure to the influence of the local priest. When the priests were expelled, as might be expected, calls came from the Tucson presidio to refigure the water allotment. Tucson's *jefe político,* Manuel Escalante y Arvizu, described the water source in 1828 as "a magnificent spring of water that gives life to its extensive agricultural lands" and argued that "legal steps should be taken to award at least half of Tucson's water to the settlers."[14]

The expulsion of the Spanish-born priests was part of the general secularization of the missions, which included legislation seeking to distribute the mission lands as private property to the Indians, as well as return to the Indians land illegally taken since 1811. Seemingly a boon to the Indians, this policy followed the typical dismal history of subsequent allotment policies. As Officer noted, "In some cases, the Mexicans paid individual Indians for parcels of land, but in others they simply took what they wanted." In El Pueblito at this time only eight Indian families remained, and they were increasingly under pressure to sell their land and hand over their water rights. In 1829 a retired presidial soldier, Francisco Herrán, bought a plot of land at El Pueblito for sixty-eight pesos from a Pima Indian, and then, when the land proved insufficient for the subsistence of his family, Herrán petitioned with the local justice of the peace for another plot, which the court promptly awarded.[15]

One result of government policies concerning settlement in the valley was the transition of the Indian population from Pima agriculturalists to Papago stock raisers and Apache pensionaries. The government had encouraged the settlement of peaceful Apaches near the presidios and this had begun the shift, complete by the late 1820s, from Pima residents to Apache and Papago inhabitants in the middle and upper basins. The Pima agricultural-ists had participated in the extensive agricultural production during the establecimientos de paz, which had overtaxed the river's flow, but since then, due to the wide-ranging effects of epidemics and cultural deficit, the Pima Indians had declined drastically in number. Near the Tucson presidio, Apache residents grew in number while the Pimas of El Pueblito dimin-ished, but the peaceful Apaches never became serious farmers in their stead. The remaining Pimas and Papagos responded to the inroads into their land and water rations by increasingly turning to mere subsistence agriculture, leaving the presidio residents with the majority of water in the basin, but with nothing resembling the bounty of the former agricultural surpluses.[16]

Indications of the declining conditions of settlement in the middle basin were clearly evident by the end of 1828. Circumstances in Tucson were severely pinched. The raids on the cattle herds—including one earlier that year—had seriously reduced the number of livestock. The only way to protect livestock was to watch them constantly. Residents pastured horses, oxen, and bulls during the day, but brought them inside the presidio at night. In Tucson the commander reported "despair" among residents, which had manifested itself in a vote to abandon the outpost for more secure environs somewhere to the south in Sonora. Escalante's 1828 report to the governor asking for a reallotment of water came in the wake of this settler unrest.

Similar conditions affected the settlements to the south. In 1829 a report on the conditions at Tumacácori indicated that the mission lands "had declined greatly in value." The diminished cattle herd of less than 500 head was "running wild in the desert" and the communal mission fields were lying fallow. The drop in production at Tumacácori since 1820 had been remarkable. At that time fewer than 200 Indians and Mexican residents had produced 160 fanegas of wheat, 12 fanegas of maize and 16 fanegas of frijoles. The mission community also ran livestock herds of 5,500 cattle, 1,080 sheep, 590 horses, 60 mules, and 20 donkeys.[17]

Just as increasing commerce had expanded production at the turn of the century, declining commerce in the late 1820s contributed to shrinking agricultural production. By 1828 the presidio commander in Tucson had ceased purchasing supplies from the local population. Rather, the commander acquired supplies from the provincial capital of Arizpe, relying on grain production from the San Ignacio Valley 200 miles to the south. Manuel Escalante was of the opinion that the Arizpe merchants were less than honest in their dealings, but regardless of the possible graft involved, persisted in the long-distance commerce. The cessation of a major source of local trade helped push residents into a subsistence mode of production. Lacking a market for the sale of goods to the presidio, the Mexican settlers and few remaining Indians of El Pueblito had no reason to produce beyond their own subsistence needs. Also the mission lands around San Xavier no longer had to support the priest and his building projects. As a result, most of the agricultural land near the mission lay fallow, and this at a time when farming at Tres Alamos, on the San Pedro River, had ceased. Escalante noted in 1830 that few of the mission fields were in production because the presidio had ceased purchasing surplus grain from the mission Indians.[18]

The penurious condition of the presidio resulted from misguided, corrupt, or foolish policies, not from a dry river. A similar circumstance also accounts for the declining productivity at Tumacácori. The 1829 report on declining conditions there had said that wheat production was down due to "Apache peril and a lack of demand," not a diminished surface flow.[19]

Tucson and Tubac officials surveyed conditions in 1831 in a census and a report requested by a national bank chartered to promote development. The census indicated that the population in the valley had once again congregated in the three military presidios. Tucson reported 465 residents, not including the friendly Apaches and other Indians, who probably accounted for an equal, or slightly larger number (thus about 1,000 residents in 1831; Zúñiga had counted 1,015 in 1804). Tubac had 303 non-Indian residents (the 1804 total population of Tubac had been 164). The presidio of Santa Cruz had 727 non-Indian residents. The total non-Indian population in the river valley was 1,495.[20] Whereas the human population in the valley had remained stable, or even increased slightly from 1804 to 1831, the agricultural production and livestock herds had declined remarkably. Since cultivation of crops and grazing of livestock use water at a rate greater than human domestic use, this constituted a distinct lessening of the demand on the river.

A prominent Tucson merchant, Teodoro Ramírez, made the report to the bank, which made clear the river's centrality to the community's existence: "The land is fertile by nature for every kind of agriculture because of the fullness of its river." The report on economic conditions in the area also accentuated the preoccupation with the Apaches: "sheep raising is not established because of the danger of the Apache enemy," and, "there are deposits of virgin gold in the vicinity of these tribes next to the Rio Salado (Salt River). If the government of the *Direccion* undertook to find them (which the enemy impedes) they could very well be uncovered and through these treasures, they could promote new settlements."[21]

The Indian raids often met no official response due to poor conditions for the presidio troops. Such was the case in 1830 when Indians attacked the settlement at Calabazas, as well as the San Pedro ranch. In the face of increasing Apache activity, many residents of the upper basin headed for Tucson. By June 1832, only twelve civilians remained in the presidio at Tubac. Nothing but the defense provided by the Sección Patriótica stood between the beleaguered settlers and the hostile Indians. Later in 1832 the militias of Tucson and Tubac gained a victory against the Apaches, and by the end of 1833 some of the Tubac residents had returned to the presidio, but the situation remained tenuous throughout the remainder of the Mexican period.[22]

Despite incessant warfare with the Apaches, subsistence activities churned along in the upper basin, and soon conflict in a typical form developed between settlers and sedentary Indians. The few persisting Pimas at Tumacácori had suffered through a drought in 1832, and then found themselves under the abusive control of José Sosa of Tubac. In 1833 the Indians charged Sosa, the administrator of the mission lands, with embezzlement and other crimes. Records of the testimony at the trial survive, but either no verdict was rendered, or the verdict itself has been lost. Two years later Sosa was again before the authorities, this time the plaintiff in the case, arguing that the residents of Tumacácori were diverting all the water in the river, which he needed to irrigate his own fields downstream at Tubac. The Tubac Justice of the Peace, Juan Elías, had found in the investigation into José Sosa's complaint that there was plenty of water, but it required hard work to deliver it to the crops. Namely, irrigation canals required regular cleaning, since vegetation on the banks and within the canal would soon restrict and eventually clog the flow of water through the acequia. The Sonoran officials reprimanded Sosa for "laziness." Elías had found that Sosa's indolence rather than a lack of water had deprived the fields in Tubac of water.[23]

The ebb tide of settlement continued through the 1830s as Apache raids made living on isolated ranches too dangerous. Vaqueros working on the ranches had been easy targets for Apaches since at least 1830. Most of the land grantees ceased living on their grants by the end of the decade. The Arivaca and Babocómari ranches, for example, were abandoned in 1835. The Ortiz family continued to work their ranch on the Canoa grant through the 1830s, but lived in Tubac. Ranching in the San Rafael Valley continued into the 1840s, until a devastating Apache raid in 1843 killed approximately thirty settlers near present-day Lochiel, and organized ranching in the valley ceased along the headwaters of the river.[24]

The surviving ranchers and their families from all over the valley tended to congregate near the presidios, primarily at Tucson. As the presidios' populations swelled with migrants from the countryside in the late 1830s, farming resumed at Tres Alamos on the San Pedro River. Soldiers from the presidio at Tucson accompanied the workers who planted and harvested the crops. The typical practice was to plant in the late spring, taking advantage of the summer rainy season, returning later for the harvest. Although the overall population in the valley was down, at Tucson the population had increased and the need for intensive agriculture returned to the middle basin. In this circumstance the Mexican population, augmented by stockmen and their families, proved unable to maximize the river's flow as Hohokam and Pima agriculturalists had done in decades and centuries past. The model for efficient use of the river was the Hohokam, but the Mexican population could not match the Hohokam's productivity. The major reason for the failure to farm more efficiently was the continuing effort to graze livestock along the river, and not just in the form of beef cattle. Even as conditions grew more and more pinched in the presidio, Tusconenses still pastured dairy cattle on the floodplain fields. In this light, the picture of presidio life at times perplexes. On the one hand presidio residents seem to have lived under siege. Presidio soldiers in Tucson escorted civilians who left the walled town to do daily chores such as to wash clothes in the irrigation ditches, or to go on forays onto the ranches to acquire meat from the feral herds of cattle and sheep. On the other hand, presidio residents blythly maintained inefficient agricultural practices, epitomized by dairy cattle consuming valuable grain and the continual bickering over wandering livestock in the Indians' cornfields.[25]

In Tucson, twin circumstances of declining production and complaints by Indians regarding the trespass by Mexican livestock caused fresh grievances of a familiar sort. In April 1836, the visiting priest at San Xavier wrote a protest to the Tucson justice of the peace. The priest asked the presidio official to enforce laws regulating the pasturage of horses and dairy cows, which were trampling and eating Indian grain at San Xavier and El Pueblito. Declining production the previous year had caused the reduction by half of the rations given to the peaceful Apaches remaining at the presidio. At this time almost 500 Apaches lived outside the presidio, and their total wheat ration was reduced to slightly more than eighty Spanish bushels per month. At eighty bushels a month, the ration entailed consumption of about 960 bushels a year. If this was half the previous ration, the Apaches had been consuming over 1,900 bushels of wheat a year, a level of production now apparently unreachable. (Recall, if you will, Manuel Escalante's report just eight years previously in 1828: presidio residents had produced a wheat crop of 2,000 to 2,500 bushels annually, "enough for all their needs.")[26]

Conditions at Tucson and Tubac continued to decline into the 1840s as more and more settlers crowded into the presidios. In 1842 the Tucson commander, Antonio Comadurán, informed the governor of the difficulties in providing supplies for an upcoming military operation. Comadurán had been ordered to provide "forty men, some cattle, and 100 fanegas of pinole, as well as a quantity of salt." Unfortunately, the agricultural production at the Tucson presidio had so declined over the years that even these modest supplies were unavailable. Comadurán wrote the governor that the community could not spare 100 fanegas of wheat to make the pinole, only a surplus of sixty fanegas was available. The presidio completely lacked salt. The presidio even lacked the cattle hides used to carry the pinole and mules necessary for transporting the supplies (the presidio had only ten mules, already appropriated by presidial soldiers). Three months later, the Tucson commander reported that the presidio had only twenty horses. In comparison, in 1804 the presidio horse herd had been 1,200 head, and the number of mules had been 120.[27]

In 1843 the justices of the peace of Tucson and Tubac made reports that indicated just how much productivity in the area had declined. At Tumacácori, the church was holding up well enough, but outbuildings on the grounds, including the convent built in 1821, had deteriorated beyond repair. Agriculture had decreased apparently to mere subsistence levels by a much-reduced population. Indians irrigated only a few small plots, and most of the mission land was overrun with mesquite and scrubby brush. Upstream at Calabazas and Guevavi no agricultural production took place, the fields completely abandoned. So meager was the production from the mission at Tumacácori, the state government in 1843 declared it abandoned, and sold the land at public auction at Guaymas.[28]

The 1843 report by Tubac's justice of the peace also gave evidence of unregulated grazing in the area. The spread of mesquite and brush onto the mission fields must have been preceded by a thinning of the grass coverage, since mesquite seedlings have difficulty proliferating when grass is abundant or cultivation takes place. The reference to cattle in the area also indicated potential overgrazing of the watershed. The mission cattle herd was unsupervised and wild "broncos" grazed indiscriminately on the hillsides.

Similar conditions existed at San Xavier. The church was in good condition, with the exception of damage to paint and surrounding structures caused by moisture and humidity. Agriculture was at much-reduced levels. The priests' fruit trees had been neglected to the point they no longer produced fruit, and most of the land formerly cultivated to support the priests and their building projects was fallow. The Indians living near the mission had apparently turned to subsistence agriculture. The report said that the houses of Indians "dotted the landscape," but few of the Indians actively farmed. These references seemed to indicate that a few Indians engaged in agriculture were routinely producing the subsistence needs of many Indians.

The report blamed the declining conditions around the missions in the Tucson Basin in part on the actions of the visiting priests (since 1828 only circuit priests had served San Xavier and Tumacácori). Specifically, Tucson's justice of the peace blamed Father Rafael Díaz, recently deceased, for selling off the mission livestock, as well as the furniture of the mission, including the doors. The liquidated livestock included oxen used by the Indians for plowing the fields. The justice of the peace presumed—in the manner of the Spanish and Mexican settlers as far back as Father Kino—that the loss of oxen or other draft animals tremendously hindered the Indians' agriculture. Of course the Indians had successfully farmed for millennia without domesticated livestock to assist them. Had the nineteenth-century Pimas and Papagos lost the knowledge and skills to undertake intensive agriculture without the assistance of domesticated livestock? The 1843 report noted the Indians' facile reversion to earlier agricultural modes of production when oxen were unavailable at San Xavier. At those times Indians planted their crops with sticks in the traditional manner.[29]

At Tucson and El Pueblito agriculture sputtered, but the report made clear that a lack of water in the river was not the cause. Tucson's justice of the peace reported that the mission buildings at El Pueblito were in bad shape, but repairable, and "the mission fields with their abundant irrigation water are cultivated by six Indians, the only ones still here, and by residents of this presidio." The Mexicans were farming without paying any rent since the priests had freed them from that obligation in 1839. The rent had been in the form of grain: "one fanega of grain for each fanega of seed they planted." The proceeds from rent payments seemed not to have been used for the benefit "either of the mission or the Indians." One example of the incomes and proceeds from the mission fields not going to the benefit of the local population is the winery that Father Díaz apparently established in Tucson in the late 1820s. The Tucson justice of the peace said that the priest took the output from his winery to his base church at San Ignacio in Sonora to the south. Tucsonenses make no reference to the production of wine in the 1830s or 1840s. So, although they had corrected the failure noted by commander Zúñiga in 1804—that Tucsonenses were missing the opportunity to cultivate grapes for wine—they were not able to enjoy the benefits from that labor until the 1850s.[30]

Further decline was reported in 1845, on the eve of the Mexican-American War. In April 1845, twenty-eight soldiers manned the presidio at Tubac, and the Tucson garrison contained seventy-eight ill-equipped soldiers. The shortage of livestock in Tucson included the soldiers' mounts, numbering twenty-eight, fifty fewer than the soldiers on hand. Not only were the horses too few in number, they were also in terrible condition, unable to withstand a journey of more than a few miles. To compound matters, the Tucson commander reported that the garrison's larder was bare and its coffers empty. He ended up begging one hundred fanegas of wheat from the residents of the presidio. The settlers were eventually paid by the local merchant, Teodoro Ramírez, who financed the purchase of the supplies by selling tobacco and cigarettes.[31]

Tobacco and cigarettes? Given the dire descriptions of conditions on the northern frontier, the sale of tobacco seems anomalous. Just how poor was the Tucson community? How circumscribed had life in the presidio become, and to what degree did the condition of the river correspond to the condition of settlements along the river's banks? As more fields fell out of production, and as livestock herds on the floodplain diminished, how did the river respond?

The Indians understood the beneficence the river could still provide. Fernando Grande, the interim caretaker of the mission lands at San Xavier, had observed in 1830 that the mission fields were largely unused due to the lack of a market at the presidio downstream, but that "San Xavier was still visited by Papagos from the villages of Santa Ana and Santa Rosa who helped the San Xavier Indians with their harvests; and by Pimas from the Gila River towns, who came in the winter when their own food supplies were low." A similar remark in 1843 had characterized El Pueblito's irrigation water as "abundant."[32]

Earlier, during the establecimientos de paz, the river had been stretched to its limit, but by the late 1820s the demand on the river had slackened, and the stream must have returned to something like its old meandering pattern. Tacit understanding of this flush circumstance appeared in the request to renegotiate the water allotment in Tucson. Escalante had noted the spring's ability to irrigate "extensive agricultural lands." Whether the water supply was sufficient to the task was not in question, the issue was who would control the production from those extensive fields. The government policies and shifts in development had resulted in less demand on the river. The increased Apache activity (joined by Papago attacks in the 1830s) reduced overall population levels in the region, both human and livestock. This result had become pronounced by the 1830s, but its effect on the river has been rarely noted by historians. Reports by Mexican officials contain few explicit references to the river's condition, but tangential references point clearly to a recovered river. The river's flow returned, but its image in the texts remains that of a meager stream barely capable of sustaining modest populations.

Although the river had largely recovered, it did not go unchanged. A caveat to this scenario of recovery is that in the upper basin the diminished cattle herds were grazing on the watershed, turning wild and increasing in numbers, and this may have impacted the river by influencing runoff patterns in the valley. Clearly the variable of human settlement had changed dramatically since 1820, but the other variables acting on the river remained in place, including weather and watershed ecology.

With the severe decline in population in the upper basin in the 1830s, a lack of water was not the circumstance that limited agricultural production. The Tubac Justice of the Peace, Juan Elías, noted a rare spring flood of the river in May 1834—more water than anyone could use. In July of that year, Elías reported to the governor about the flood, or floods on the river earlier that year. Elías explained that "our presidio is without a wall," and the winter rains of 1833–1834 had been so heavy that they had "changed the course of our river, placing our water almost beyond reach, especially if we are under siege by the enemy Apaches."[33]

The Spanish and Mexicans had long recognized the normal pattern of precipitation in the region: monsoon-type downpours in the summer (*las aguas*), and a winter season of slow, steady rain (*equipatas*).[34] The first priest assigned to the mission at Guevavi in 1701 had been delayed by monsoon rains that had flooded arroyos and made the roads muddy.[35] By 1833 the roaming cattle herds had perhaps begun the process of denuding the hillsides throughout the upper basin so that by the time the winter rains came there was scant vegetation on the hillsides to impede the rush of water down to the river channel. Many factors go into the production of floods in a semiarid region. The most prominent factors are the cycles of rainfall and the impervious nature of desert soil. Summer deluges often create flash floods in their normal occurrence, facilitated by the sparsely covered desert soil, often made up of highly alkali caliche.[36] In addition to these more-or-less normal floods, climatological factors can play a role when above-average rainfall occurs, perhaps due to El Niño episodes.[37] Other factors mitigate against flooding, such as thick coverage by grasses as naturally occurred in the upper basin, limiting runoff and facilitating infiltration of precipitation. The floods of 1833–1834 may be the first incidence of floods in the valley influenced by overgrazing.

This is not to say that overgrazing itself was a process that took place somehow outside of "nature." The long environmental history of the river certainly included examples of such processes. Many grazing species had inhabited the valley over time, proliferating to unknown levels, at times denuding the watershed in the absence of efficient predators, trampling stream banks and destroying riparian plant species as they daily migrated to water sources. Ancient, extinct species of mammals no doubt witnessed floods on the river, periods of arroyization, infill, cienega formation, and course changes, long before the first humans arrived in the valley. The floods of 1833–1834 may be simply the first floods on the Santa Cruz River with human society as an actor in the process.

In the Tucson Basin residents at the turn of the nineteenth century had engaged in nascent commerce to such an extent that the river no longer provided enough surface flow for the valley's residents. By the late 1820s the residents once again appeared unable to sustain themselves, but this time due to drastically reduced levels of agricultural production. Through the 1820s and 1830s, political, economic, and social factors pushed the most-efficient farmers in the Tucson Basin into subsistence patterns, with no surplus available to sustain the presidios when trade, not the river, dried up. As long as trade routes and transportation links to the south remained open, and if the government's finances allowed it, the presidio could obtain the supplies it needed. But with the increased hostile activity and the insolvency of the government, that trade melted away, and the Tucson presidio suffered shortages, as arable land nearby lay fallow. The Mexicans had legal claim to half the water in the basin and they used this, and more, to support agriculture and their livestock herds; the Indians had legal claim to the other half, and used much less than this to support themselves and their kinsmen.

Why didn't the Mexican population simply expand agricultural production in the basin to match their needs? As Escalante noted in 1828, the Mexican population adopted a similar subsistence mode of agriculture due to the lack of a reliable and consistent presidial market. But also, Hispanic farmers could never match the efficiency of their Native American neighbors. As Meyer explained, the Hispanic effort to duplicate the Hohokam irrigation systems met with remarkable failure. In another example, the Europeans expected to prepare their fields by plowing. Oxen or some other draft livestock were used for this purpose. As mentioned previously, Fernando Grande had reported the decline in San Xavier acreage under cultivation because the circuit priests had sold off the oxen the Indians had used to plow their land. In Grande's mind there was a firm connection between livestock and agriculture: no oxen, no plowing, no farming. And the assumption was not only Grande's, it seemed to be characteristically European. Friar Bringas in 1795 had encouraged the Spanish authorities to provide the new Papago residents at Tucson with "oxen and tools" so that they could begin agriculture in the valley. Kino had held the same views. Intensive agriculture utilizing digging sticks simply fell outside the realm of Eurocentric experience.

During the dark days of 1828 Tucson residents had voted to abandon the presidio. Manuel Escalante had explained to the state governor that the settlers' despair was the result of declining prospects and spreading impoverishment: "The major factor in their decision is certainly the poverty of our military economy, upon which their civilian economy is totally dependent." As part of the solution for this problem, Escalante recommended that, "The military should be commanded to give first preference to the private enterprise of Tucson's civilians and should be forbidden to look for better prices elsewhere in provisioning their troops with food and other necessities." Restoring the commercial economy held a prominent place in his prescription for the ailing presidio, and Escalante also described the renegotiation of the water allotment between El Pueblito and the presidio as an absolute necessity. Escalante assured the governor that restoration of the presidio's previous level of prosperity would soon follow the implementation of his recommendations.[38]

Escalante possessed the memory of past prosperity, or perhaps simply possessed the imagination and perception necessary to see the river's possibilities. The river always appeared small, meandering, "gentle," but capable of irrigating "extensive" fields. When acreage under cultivation declined in places by as much as 80 percent, when the human population declined from thousands to hundreds, and when livestock herds diminished from tens of thousands to dozens, what happened to the river? As the geology of the region dictated, the river still disappeared beneath the sand as it left the middle basin, unless augmented by floodwaters from the Rillito and Brawley washes. Within the middle basin, the river remained constant, steady, available. Natural processes of surface water hydrology determined the river's flow. The 1843 report on conditions in the basin mentioned as one cause of declining agricultural production at San Xavier "a recent lack of the necessary water, in turn due to neglect of keeping the river water moving and, instead, allowing it to gather in stagnant pools along the side."[39] Clearly a cienega had begun reforming near the mission in the vicinity of past cienegas long since tapped out of existence.

The final stage in developing this circumstantial argument of a returned river is to observe what happened in the valley when a market for agricultural surpluses returned in the late 1840s. Did agricultural production in the valley spring to life as it only could if supported by an available, gently flowing, river? The answer to this question comes in the next chapter as Americanos arrive in the valley, anxious to trade with local residents.

5

Americanos
1846–1850

ACCORDING TO THE OFFICIAL REPORTS, conditions at the presidios remained pinched and poverty-stricken as the war between Mexico and the United States began in 1846. The primary American objective in the war was expansion, and the acquisition of California the main goal. The region of the Santa Cruz River valley, on the other hand, was not thought to be particularly valuable or desirable, and when the war ended, the Santa Cruz Valley remained part of Mexico, not part of the Mexican cession of territory to the United States. But the river valley did serve as a thoroughfare on the way to California, and this was its main significance during the war. It was an area to be gotten through. Into this situation arrived the Mormon Battalion, detached from General Stephen Kearney's Army of the West to traverse the northern Sonoran frontier. Some of the observations made by the Anglo soldiers put the official Mexican reports of dire circumstances in doubt, or at least provide another perspective on conditions in the river valley. At the very least the arrival of these Americanos, followed within a few years by the flood of gold-seeking forty-niners, tests the notion that the river had recovered and was capable of producing much more than the gloomy reports indicated.

Evident from the Anglo accounts of the river valley is the continued archaic, or premodern, perception of the river limited to its surface flow. Although the Americans brought a heightened interest in commerce and trade, their vision of the prospects of the river valley remained fixed on the water trickling before their eyes. These early Anglo accounts came before windmills proliferated throughout the West, and the difficulty, cost, and reliability of digging wells and lifting groundwater to the surface remained beyond the reach of any reasonable endeavor. No doubt the American soldiers and sojourners would have bridled at the notion that their vision of the river linked them more closely to the prehistoric Hohokam and Pima inhabitants of the valley, rather than their modern American relatives of the late-nineteenth century, but such seems clearly to be the case.

The Mormon Battalion was not the first group of Anglo-Americans in the Santa Cruz Valley, but the first Americans in the region—fur trappers in the 1820s—left few records of their impressions of the area. With the close of fur trapping in the region as the number of beaver decreased and Indian hostility increased, American interest in northern Sonora ended until the Mexican-American War twenty years later. After this interim of distinct apathy, military and strategic interests brought Anglos back into the region. General Kearney ordered the Mormon Battalion, commanded by Col. Philip St. George Cooke, to take a southerly route through northern Sonora to the Pacific coast and California, where the general planned to provide coastal shipping to transport the troops to wherever they were needed—perhaps the Sacramento Valley. In the meantime, Kearney would take the main army directly from Santa Fe to California.

The battalion was composed of about 500 men, several of whom kept journals during the expedition. They entered the region of the Pimería Alta from the south and east, near the San Bernardino Ranch in December 1846. There the soldiers encountered "numerous herds of wild cattle" that could be hunted like buffalo. The meat, according to one soldier/chronicler, Henry W. Bigler, was "the sweetest beef I ever ate." Estimates of the number of cattle roaming the grasslands varied, but the herds were no doubt large and impressive. On December 2, Bigler wrote "It is supposed we saw four or five thousand head of wild cattle." Three days later, Robert S. Bliss reported camping at a spring "where the best judges think 10,000 head of cattle come for water." Bliss went on to give a casual indication of the rapacious ethos of the group: "we kill all we want and more than we need." He also echoed the positive assessment of the taste of wild beef: "their meet is fat and tender, the best beef I ever eat [sic]." To the amazement of the soldiers, the numbers of cattle, antelope, and wild horses increased as the column traveled downstream to the north, through the San Pedro Valley.[1]

The battalion had fairly sighed with relief when they reached San Pedro River valley following a dry and difficult journey from the Rio Grande Valley. One described the San Pedro Valley as "beautiful," and another called the river "a fine bold stream." The grazing wildlife provided plenty of meat, and the stream yielded trout of impressive size. Bigler wrote,

"Some of the boys said they were salmon trout," and Colonel Cooke noted in his journal, "Salmon trout, eighteen inches long, were caught."[2]

Near the area of Tres Alamos the battalion left the San Pedro River and traveled west toward Tucson, passing a group of Mexicans distilling mescal along the way. The column entered Tucson the next day and found the people there open and amenable to their arrival, although the presidio troops and most of the residents had earlier moved to San Xavier to avoid any confrontation. The battalion had expected about 500 inhabitants in Tucson, but found only about 100, mostly "old men and infirm, with a few children." The Mexicans welcomed the Americans into the presidio, offering them water as the soldiers passed by. As James Officer noted, the gift of water was "perhaps the most hospitable gesture a desert people could make."[3]

The trading was generally cordial, though the locals were "close in their dealings." One soldier reported trading clothing for "beans and flour." The agricultural possibilities in the area were obvious to Bigler, who seemed to express a nostalgia for farm life: "It looked good to see young green wheat patches and fruit trees [this in December], and to see hogs and fowls running about, and it was music to our ears to hear the crowing of the cocks. Here are the finest quinces I ever saw." The soldier also noted the presence of "two little mills for grinding grain and run by jackass power, the upper millstone moved around as fast as Mr. Donkey pleased to walk."[4]

The soldiers' reports hardly conform to official characterizations of the presidio's dire circumstances. This is not to say the battalion walked into the lap of prosperity, but the subsistence pattern of agriculture that had existed in the area since the late 1820s apparently included a broad range of agricultural products, although limited in abundance. The battalion acquired during its one-day layover in Tucson three bushels of salt, helped themselves to some portion of 2,000 bushels of wheat found in the presidio, and traded for flour, meal (perhaps pinole), tobacco, quinces and pomegranates. The greater potential of the valley seemed obvious to the American newcomers.[5]

As the Mormon Battalion moved on, the settlers in Tucson and the valley returned to old patterns of Indian wars and retrenchment. The Mexican and peaceful Indian population continued to concentrate in the Tubac and Tucson areas, as evidenced by the number of baptisms and their locations in 1847. Meanwhile the effects of the war began to reach even remote outposts on the northern frontier. In February 1848, the Sonoran government was forced to suspend the "bimonthly wheat ration" to Tucson. The U.S. Navy had blockaded the port of Guaymas, which had reduced the Sonoran government's revenue, and thus its ability to provide the rations. With the approach of spring, the Tucson commander, Captain Comadurán, reported that starvation faced many residents. The short-term solution was to round up some of the wild cattle roaming the valleys. A group of fifteen presidio soldiers and settlers attempted to obtain beef at the old Babocómari Ranch, but were attacked by hostile Indians and all were killed. Nine of the soldiers left widows in Tucson who later petitioned to the Sonoran government for their entitlement of military pensions.[6]

A census of the frontier settlements in 1848 indicated the degree of concentration of the population in Tubac and Tucson. Tubac had a population of 249, with an additional 200 peaceful Apaches nearby. Tucson's population was 760. Considering the number of settlers and Indians scattered throughout the river valley in earlier times, this concentrated population in Tubac and Tucson represents a distinct drop in overall population. For the first time since 1804 the non-Indian population of northern Sonora had fallen below 1,000.[7]

The doom and gloom reports continued from government officials. In the late summer of 1848 Apaches attacked and besieged the Santa Cruz presidio. None of the residents were killed, but the Indians made off with "60 cattle, 10 horses, and three mules." Soldiers got some of the livestock back, but could not defeat the Indians in battle. After the attack, the presidio commander, Captain Francisco Villaescusa, declared that the Apaches were "impoverishing" the settlers and that many of his troops were deserting. In another example of dire predictions and observations, the Sonoran governor, Manuel María Gándara, circulated an official woe-is-me: "All is misery, and the government finds itself lacking even the capacity to make use of the few available resources to ward off the most serious ills."[8]

Once again, historians are fortunate that just after these gloomy descriptions by the Mexican officials, a group of American soldiers traveled through the area recording their own observations. The two perceptions offer an interesting contrast. An army detachment commanded (rather drunkenly) by Major Lawrence P. Graham traveled through the Santa Cruz Valley on its way to Los Angeles from Monterrey in 1848. The purpose of the military expedition was to transfer forces from Mexico to the thinly stretched garrisons of U.S. troops in California. The journey took place after the war was over, and the Americans entered U.S. territory once they reached the Gila River, which was the international boundary established by the Treaty of Guadalupe Hidalgo. The settlement of the war had left the entire Santa Cruz River valley under Mexican sovereignty.[9]

As with the earlier Mormon Battalion, Graham's column entered the region from the east at the old San Bernardino Ranch. From there the soldiers marched west to the San Pedro Valley, but where the earlier group had followed the river north, Graham continued heading west, around the south end of the Huachuca Mountains, and into the Santa Cruz valley at Santa Cruz on October 14, 1848. Lt. Cave Johnson Couts kept the most complete journal of the trip, and he confirmed that the presidio had been attacked by Apaches three weeks earlier. Couts wrote that the Indians had "carried off all their animals but one single mule, and all their clothing."[10]

Nonetheless, while in Santa Cruz the soldiers were able to trade for corn, chicken, and pork. Shortages no doubt existed, but the surplus available for trade speaks of a less gloomy circumstance than commander Villaescusa had indicated. Couts noted that the Mexican women came into the visitors' camp regularly to trade for some essentials: "sugar and soap, soap and sugar, is the only cry." Obviously the Santa Cruz residents were not flush with surpluses of all variety and suffered from the typical shortages of a frontier settlement, but neither were they destitute and starving. Couts reported: "We found plenty of corn, of a good quality small ear, very hard and sound, commonly called *hominy corn* and instead of $10 or $15 per fanega as expected only $3.00." The unit traded for 100 fanegas of "shelled and sacked" corn, and then filled up the wagons with as much corn on the ear as they could carry. Couts said Santa Cruz had two mills, but only one was operating when the soldiers arrived in town. The old miller was producing "one fanega, or two and a half bushels of corn per day." One of the soldiers in the troop made repairs on the mill (or perhaps repaired the second mill), which then increased production to twelve fanegas per day, and produced better-quality meal and flour in the process. In return for the work, the "mechanic" received $30 from the miller, which apparently stripped the entire community of ready cash. But the townspeople were acquiring more currency with each day the American soldiers lingered. Graham's command stayed in Santa Cruz four days, during which trade continued more-or-less nonstop. Couts reported, "Probably no little town ever furnished as many pigs, or shoats and chickens, as did St. Cruz, the past four days. Everyone [the soldiers] has as many as they can lug off tomorrow and have eat[en] nothing else since we've been here. Shoats $2.00,

Chickens 2 bits or a dozen per pound of sugar." Clearly, the river continued to provide subsistence in the ancient pattern, but curiously, given the official reports of grim and abject poverty, residents had produced modest surpluses by the time the soldiers arrived anxious to trade, and this only three weeks after an Apache raid.[11]

The Americans left Santa Cruz on October 19, 1848, following the river out of town. Couts remarked, "It is a beautiful little stream, passing through the mountains lined on either side by a large growth of Cottonwood, from which extends a small but magnificent valley." They passed "several deserted as well as inhabited ranches," and an operating gold mine near the old mission at Guevavi. The Mexican miners explained to Couts that "The Apaches are so numerous and severe . . . that the work only goes on at intervals, never over two weeks at a time."[12]

The next day the Americans made steady progress past Tumacácori, and into Tubac, both of which Couts described as primarily Indian villages. The last water north of the Tubac presidio before San Xavier was at the site of the uninhabited La Canoa Ranch, and that is where Graham's column spent the night of 23 October. Before leaving the river where it sank into the sand, Couts summarized his impressions of the valley:

We knew nothing of St. Cruz or the little valley which we came down from there until it was staggered upon. The *corn* here saved us, for not a bite of grazing worth consideration did we find from there through. About a month *earlier* the grazing from St. Cruz to Goibabi [Guevavi] was good but the frost in the mountains had destroyed it entirely. From Goibabi to Tucson it is *never good,* being mesquite growth and chaparral. On the other side of the mountains, and to the south and west of our course, down the Rio St. Cruz there must be a very fine and healthy country—the climate most delightful and suited to the growth of all and every plant and fruit which is found anywhere in the Republic. Sonora is regarded as the wealthiest, the first state of the Republic, and *this* the *garden spot* of Sonora. Figs, Grenadas, and other fruit were brought to us at St. Barlora [Santa Barbara]; the finest grenadas that we had ever seen, some four inches in diameter. Pumpkins and melons grow in a wild state on the St. Cruz River. We passed several patches; and the whole command supplied themselves with pumpkins of a most elegant quality and large size. Melons not good. These are not the spontaneous growth of the soil, but *indigenous,* have evidently sprung up from seed which had been scattered by some party, and multiplied, from year to year, the winters not being of sufficient duration or severity to destroy them.[13]

On 24 October the soldiers made the easy traverse to San Xavier. The Americans were impressed with the mission, and looked forward to arriving at Tucson, which they expected to be a prosperous and bustling town. Their expectations had been heightened traveling down the river valley, and as a result they were generally underwhelmed with the old presidio: "no *great deal* after all."[14]

In Tucson the soldiers received the same sort of friendly treatment that the Mormon Battalion had received two years earlier: "[they] think Americans are *every thing.*" The major difference in the encounters turned out to be the weather: cold and rainy in October. Captain Comadurán "was nearly frozen" waiting for the drunken Graham to appear for a formal greeting. The soldiers spent only one day in Tucson, and according to Couts the rain started "about 6 o'clock, and continued nearly all day, very cold and chilly."[15]

Couts' description of Tucson was brief. He made reference to "San Agustine," apparently referring to the old Pima village of El Pueblito, which he indicated was abandoned, "a complete *dog town* covered with old broken earthenware." It was near this site that the river "disappears into the sandy desert." Couts also mentioned the grist mills in Tucson: "Every house in Tucion [*sic*] furnished with a Buro [*sic*] *flour mill,* and are kept going incessantly, probably grind a half bushel of wheat in 24 hours." Henry Bigler of the Mormon Battalion had reported only two mills in Tucson in 1846. Was Couts exaggerating, were the earlier observers simply less observant, or had the local residents expanded their productive capacity due to an increased demand? Given the relative simplicity of the mills, it seems reasonable to expect that they could be brought on line or retired with relatively little effort, as long as burros or other livestock were available to turn them.[16]

The next morning, 27 October, Graham's men left Tucson heading for the Gila River villages of the Pimas. Couts had stated that the river went underground in the Tucson Basin near El Pueblito, but the column encountered "*very muddy water*" in pools near the confluence of the river and its tributary, the Rillito. This was the last available water until the Gila River, except for the pools and puddles created by rain showers during their journey. Couts recognized the good fortune to have rain during that stretch of their journey, "What a kind Providence!" From the Gila River villages, Graham's column followed the Mormon Battalion's route to California.[17]

The American soldiers on two occasions provide verification for the subsistence pattern of agriculture that had been in place in the valley since the late 1820s. Their expectations for a thriving Tucson give an indication of their appraisal of the potential of the river valley. This in part led to the American's prejudiced opinion of the Mexican soldiers and residents of the valley. The Americans did not understand the history and circumstances that had led residents to exist in such a bare fashion. To the newcomers, a river of clear potential was flowing wastefully into the sand.

As the American soldiers left the valley, once again the old patterns of life and death resumed. In December 1848 Apaches attacked Tubac and Tumacácori, resulting in the abandonment of both settlements. The Sonoran government urged the Tucson commander, Antonio Comadurán, to reoccupy Tubac with twenty soldiers, but Comadurán replied that his men were in too pitiful a condition to undertake the journey: they lacked shoes and the weather was freezing.[18]

The former residents of Tubac and Tumacácori had scattered throughout the frontier region, many settling at San Xavier and Tucson, where they stretched the supply of rations on hand. The Tucson commander reported that some of the "refugees" had moved to the "mining community of Soni," and others had headed for the California gold fields. He then closed, with typical pessimism: "The only persons left to die of hunger or to fall victim to the Apaches would be those with large families who could not undertake the journey [to California].[19]

The raids on Tubac and Tumacácori were but two examples of the Apache onslaught at the end of 1848 and into 1849. Officer characterized these attacks as "the most devastating series of Apache raids ever experienced in the region." The raids in part resulted from increased vulnerability due to the depopulation of the presidios and frontier towns as local residents joined the exodus to California. The raids also encouraged the out-migration, convincing some of those who initially had hesitated to pack up and join the gold rush, finally to leave. As a result, the frontier settlements approached total collapse.[20]

Once again the gloomy reports by Mexican officials preceded a new wave of Anglo observations, this time provided by the parade of forty-niner caravans following the Gila Trail through the valley on their way to

California. Overall, the accounts describe the transition from subsistence agriculture and modest surpluses in the valley to increasing surpluses available for commercial market. Many of the travelers' accounts remarked on the beauty of the river valley, being much taken by the green riparian ribbon of a gently meandering river in the midst of the semiarid region. Since the chronicles are somewhat repetitive, I will give the reader but a few examples and then summarize the remainder.

The first Anglo group into the area heading for California in 1849—although it seems they were not pursuing riches in the gold fields—was a party of twenty-five headed by the explorer John C. Frémont. Frémont's expedition passed through the valley in the early spring, but Frémont made few references to conditions in the valley. Nonetheless, his cryptic journal entries indicated a few positive impressions: "The Apache visitor—Santa Cruz. The Mexican and the bunch grass. Follow down the Santa Cruz River—Tucson. Spring on the Santa Cruz—peach orchard [at Guevavi, Tumacácori, Tubac, San Xavier, or San Agustín?]—the ruined missions. River lost in the sand. The grass field and water at foot of the hills. Reach the Gila River."[21] Wherever Frémont encountered the peach orchard, the presence of fruit on the trees never failed to impress the Anglo travelers with the real and potential fecundity of the river valley.

Next in the area to leave an account of their journey, and perhaps the first group of genuine forty-niners, the Carson Association of New Yorkers entered the valley in April or May 1849. Harvey Wood kept a diary during the trip, and he reported passing many deserted villages, "caused by Apaches making a raid on the place, killing a few of the inhabitants and helping themselves to stock or anything else they fancied." Little detailed information on conditions in the valley appear in the diary, but John G. Goodman III, the diary's editor, characterized Tucson as a community which produced little agricultural surplus: "an ancient adobe town of roughly 500 inhabitants . . . so unimpressive that the overlanders mentioned it primarily as a place where they could obtain a small amount of food."[22]

In May 1849 a group from New Orleans traveled through the region. A correspondent from the *Daily Picayune,* John E. Durivage, accompanied the expedition. The reporter's account is descriptive and a bit hyperbolic at times, as the journalistic style of the time required. It is also representative of many forty-niner journals in its evaluation of the river valley's potential. As had become the common practice, the group hunted wild cattle near the San Bernardino Ranch before arriving at Santa Cruz on May 23: "About twelve miles further we turned the brow of a hill and were gratified with the sight of a small town embowered in trees, in a very pretty valley, six miles off." The town proved to be a disappointment as they approached: "To our regret, we found the town old, dilapidated, and the poorest of the poor. There was no such thing as a store in the whole rancho." Durivage was disappointed in the offerings of Santa Cruz, not because the area was inherently poor—or more to the point, dry—but because the group from New Orleans had arrived a few days too late. Another group of forty-niners—perhaps Wood's group, or a group of Mississippians traveling at about the same time through the area[23]—had passed through Santa Cruz a few days earlier, and "had skinned the place of flour, sugar, and bread completely." Trade goods remained available, however, just not easily or casually at hand. Durivage described the ordeal and expense of obtaining supplies:

> I endeavored to procure some flour, but for sometime in vain; but at length bribed the miller with a dollar to sell me 150 pounds for $10. The flour belonged to his customers, who had left their grain behind them to be ground. . . . Whether the owners of the grain ever were paid, I cannot tell. . . . The Indians have impoverished this town as well as all others; the inhabitants are all poor as church mice, but yet you cannot procure anything for gold. Silver is the circulating medium—gold will not pass current. Judge of my hunting through the entire burgh for sugar, for two hours, and only obtaining two pounds at 75 cents a pound. . . . I bought half a dollar's worth of bread by hard begging, and representing myself as a true Christian and Catolico, a bit's worth at a time. At last the women, as if unable to resist temptation any longer, gathered up their baskets of bread and fled like a flock of sheep.[24]

Durivage's companions camped near the town "in a fine green field, bordered by cotton-wood trees." Once again, the meager supplies and difficulty in trading in the vicinity crept into his narrative: "Pork here is pork, and the old adage of too much pork for a shilling would never hold good here. I endeavored to buy a small chunk of a grunter, but was run off by the price—$5!"

Durivage commented on "the beauty of the valley" and the "exceedingly fertile" countryside as the group left Santa Cruz and followed the river downstream. Near the old Santa Barbara Ranch, the group tried to pan out some gold, to no avail, having seen "particles of mica" glinting in the sun. Farther downstream the group passed Tumacácori and Tubac (Durivage misidentified these communities, as well as San Xavier, in his narrative), reporting the sites to be "very finely situated" and "well built." The first settlement encountered (Tumacácori?), "was deserted by the proprietor last February, after having been attacked by the Indians, and in his flight all the fields of wheat and corn were left standing." Also in a typical pattern, the group spent the night near the site of La Canoa Ranch, where "the river sinks into the sand, and only appears again at intervals for many miles." Durivage reported that the river's course between La Canoa and San Xavier, although dry, was marked by cottonwoods "frequently in sight." Clearly, the water table remained near the surface, even through the stretch of dry channel between the upper and middle basins.[25]

At San Xavier and Tucson, once again Durivage was impressed by the potential of the valley, but underwhelmed by the realities of community life. He reported that the church at San Xavier was in good shape, although "the smell of mold" filled the "old cathedral." The valley remained lush and full of potential, but Durivage recognized that the sedentary Indians had adopted a strict subsistence routine: "The lands in the vicinity are rich and fertile in the extreme, and well timbered with mesquite. A few cattle, horses and mules are raised by the aborigines, and just corn enough for their own consumption." Tucson was "a miserable old place, garrisoned by about one hundred men." Provisions were limited in the presidio, but Durivage was able to acquire "flour and a small quantity of corn." Also he noted that the old mission gardens near Tucson "were well stocked with fruit. The whole valley is exceedingly fertile." The group set up camp one mile from the presidio, and remarked while there: "The camp has been filled all day with Mexican women and Indians, all eager to traffic and anxious to buy needles and thread. A few good purchases of mule flesh were made." Clearly, the American presence was stirring commerce to life.[26]

The reputation for difficulty of the journey from Tucson to the Gila River had taken on epic qualities. Durivage's group rested prior to the eighty-mile trip, and the reporter heightened the drama of the journey by remarking, "On the whole road there is neither grass nor water, and it is a serious undertaking to cross it. The springs, wells and runs which existed when Col. Cook passed it are now dried up, and it is a severe trial for the animals to cross it. Gourds were in great demand all day, and ten times their value was paid for them." Durivage's group left for the Gila early in the morning, June 1, 1849. They traveled until two a.m. the next day, June 2, rested a few hours, then moved on, arriving at the river at two p.m. The reporter and his companions were thoroughly parched by the time they reached the river. Durivage reported the scene as the animals sensed the nearness of water: "within half a mile of it, my tired mule smelt the running water. She pricked up her ears, gave one long bray and struck a bee line for the Gila, directly through the thick chaparral. . . . There was no checking their impetuosity; some of their riders were left hanging in the branches of the trees, some were thrown and some were pitched headlong into the water." The ordeal was over, with few kind words for the adventure: "I consider the crossing of this jornada of eighty miles an era of my life, and shall never forget it to the day of my death. It came very near proving the cause of my stepping out from this sublunary [sic] sphere, and that I ever did arrive at the Gila in safety was almost a miracle."[27]

By late spring a more or less continuous train of forty-niners was snaking through the valley. At times a day or two separated the groups of gold-hunters, and at other times the travelers bunched up around the trading sites of Santa Cruz and Tucson. The residents of the valley soon began to take note of the succession of caravans bringing visitors with a keen desire to trade. The spring season was ripening fruit and bringing other produce to harvest, and increased labor was being applied grinding wheat into flour and preparing pinole for sale. As trade became constant, residents of the valley took advantage of their seller's market and began driving hard bargains for food supplies.[28]

The growing seasons determined what food was available for consumption and trade. The wheat and corn harvest could be stored for some time and doled out piecemeal, but fruit seemed to be consumed as soon (and sometimes just before) it ripened. In the late spring, peaches, pears, and quinces were available at Tumacácori and Tubac. Later in the summer, wild cherries ripened, along with apples, grapes, and pomegranates. In the autumn, beans, peas, squash, and pumpkins became available. Cattle, chickens, sheep, and pork could be had throughout the year, but often at high prices.[29]

In exchange for food supplies, forty-niners traded currency (mostly U.S. dollars; however, the Mexicans preferred silver) and merchandise. Several chronicles reported a brisk trade for jewelry, handkerchiefs, and "all articles of merchandise and especially dry goods." By late summer, prices—in U.S. dollars—had stabilized within a context of local shortages and seasonal availability. Flour ranged in cost between $6 and $7.50 per fanega, and cattle sold for about $12 a head. Occasionally the travelers had the opportunity to splurge on delectables. In late September a group extolled the marvels of "cakes of sugar," which sold for twelve and one-half cents for each half-pound cake.[30]

Many forty-niners remarked on the inconvenience of trade in the valley, this primarily because the Mexican communities had no mercantile stores. When a caravan arrived outside Santa Cruz or Tucson, residents would come out to the camp with goods to trade, or the travelers would walk through town, going door-to-door asking to trade for food supplies. The process could be made easy or difficult, depending on the desires of the residents. Clearly, by midsummer, residents felt no need to accommodate every whim of the customer, and when faced with particularly rude or obnoxious behavior by the Anglos, simply closed up shop. Ethnocentrism was common among the forty-niners; Officer described the American behavior as oftentimes "arrogant."[31]

The ease or difficulty of trade affected the forty-niners' evaluations of the river valley. In general, the travelers remarked on the extreme fertility of the valley and the overall beauty of the scene. But evaluations of settlements and residents varied between grudging admiration and outright racism, and these appraisals often sprang out of the trading relationship. The pejorative stereotype of Mexicans as "lazy" can be traced back, in part, to these forty-niner chronicles.[32] The Americans could never quite understand the realities faced by Hispanic residents of the valley, especially while the Apaches allowed the caravans to pass through the valley unmolested. To the Anglo travelers, the valley was capable of producing huge surpluses with a modicum of work, but the residents seemed disinclined to apply the necessary labor to the task. Of course the residents had long since come to understand that large surpluses only encouraged the Apaches to come down and take it, so their subsistence habits had become cautious and constrained. The river allowed quick bursts of production, however, and accounts make clear that surpluses were available in the valley, just not plainly apparent or obvious.

Generally the hostile Indians did not attack the caravans of travelers, but they maintained their pressure on Mexican settlements. For example, in a July raid on Santa Cruz, the Apaches "plundered livestock, killed a visitor from Altar and kidnaped [sic] three residents." After the raid many residents left Santa Cruz—continuing the ebb and flow of settlement in the face of Apache attacks—so that a group of forty-niners from Ohio traveling through the valley in August reported the town "poor and almost deserted." The ranches between Santa Cruz and Tucson had been abandoned due to Apache raids, and the forty-niners often remarked on the "flourishing conditions," including ripening fruit on the trees, that seemed to promise great prosperity for anyone willing to work the land.[33]

Another factor shading the accounts was the climate of the region. The summer monsoon rumbled to life in July 1849, and the rain, although raising up "mosquitos immense," brought welcome relief from the heat in the valley. The dry jornada from Tucson to the Gila became much less of an ordeal after the monsoon showers. As one traveler reported, "In the dry season there is not a drop of water and but little or no grass. But luckily for us, there had been several rains, and we found holes of water all along the road. And there was considerable young grass springing up, which helped amazingly; but it was only in spots that we found this."[34]

The monsoon had brought the valley's grass to life, and the forty-niners' livestock benefitted greatly. The wild cattle in the region also benefitted, but the herds on the San Bernardino Ranch had been winnowed to such an extent by constant hunting that one diarist, Cornelius C. Cox, from Texas, remarked, "Tis strange that no other class of cattle are found here—I am of the opinion that the stock will eventually *run out* [Cox's italics]."[35]

On the eve of the forty-niner migration, Mexican officials had reported dire shortages in the valley and settlers on the edge of starvation. Clearly, the arrival of a market had created the impetus for surplus production beyond mere subsistence needs, and as a result the productive capacity of the river came to the fore. This is significant for a study of the river in that it adds weight to the view that during the earlier periods of slackening demand, the river had rebounded. No one remarked explicitly about this circumstance, but the evidence of quickly increased production indicates that the river was there, flowing and available, and as a result, a modest level of prosperity returned to the valley. Officer suggests that the trade in Tucson in 1849 indicated that commerce and prosperity reached levels "not previously known." For some inhabitants of the valley, however, the lure of the California gold fields proved greater than the desire or ability to "mine the miners," and many left the valley. And of course, old troubles continued to push people into seeking better opportunities elsewhere. Apaches still targeted the Mexican communities, and increasingly the Anglo travelers.[36]

Had capitalism finally brought the river back to life? Of course not. No attempt to valorize the trade initiated by the Anglo gold-seekers is intended here. Eventually trade and commerce in the valley would profoundly affect the river's circumstance, but at this time the accounts of trade and haggling for goods simply indicates in a tangential fashion the status of the river. It was there, constant in its surface flow, recovered from the demands placed upon it early in the century during the establecimientos de paz.

6

New Borders
1850–1856

THE TROUBLED HISTORY OF HISPANIC settlement in the region continued in the years following the Mexican-American War. Indian raids, a cholera epidemic, and shifting governmental policies and administrations fostered both development in the river valley and periods of retrenchment. Once again, in the face of official reports expressing despair and gloom, Anglo-American travelers in the region added another description of the region to the historical record that sometimes supported, but often challenged, official pronouncements. In particular, were the observations of the valley's moist conditions by surveyors and engineers, which presaged the shift to a modern perception of the valley's water resources. The river continued to flow regardless of the social and political events taking place on its banks. One event of tremendous significance for the people in the valley, although of dubious importance for the river itself, was the Gadsden Purchase in 1853, which transferred control of most of the river valley to the United States government, and in a single political act transformed hundreds of Mexican nationals into American citizens. The new boundary set out by the agreement intersected the river twice, creating a geographic anomaly claimed by the Santa Cruz River alone: the only river to originate within the United States, flow out of the country, and then return.

A series of Apache raids in December 1849 and January 1850 disrupted the burgeoning prosperity of the valley. Although no one was killed in the December 1849 raid at Santa Cruz and losses of livestock were minimal, in Tucson Apaches made off with a herd of cattle, and in Tubac the raids caused a new exodus of refugees into Tucson. The Mexican troops in Tucson were unable to follow the raiders because the soldiers were, in typical fashion, short of horses. The civilian population may also have been short of horses at this time because of the selling of livestock to the forty-niners. Another large Apache raid on Tucson occurred almost a year later, in December 1850. Like bookends, these two large raids frame a series of lesser raids that occurred throughout the year in the valley.[1]

In the midst of continuous Apache hostility during the summer of 1850, the Mexican government instituted a new policy on the frontier to bolster the defense against the Indians, and just in case, against any new American incursion into Mexican territory. The new policy replaced the old presidio system of settlement with "military colonies." A stipulation of the new policy stated that the colonies would no longer receive their ration of grain directly from the government, but rather the government would contract with private parties for grain to dispense to the peaceful Indians, and the colonies' soldiers would be expected to grow their own. Hopefully the new policy, by creating a local market for agricultural production, would attract colonists to the frontier who would both stimulate development and participate in the defense of the region. The military colony of Tucson, initially and theoretically, was expected to produce enough grain to supply 150 men for an entire year.[2]

The new policy also required that local officials find vacant land for the settlers. In Tucson, the recent refugees from Tubac added to the list of those seeking land. During the period of slackened demand beginning in the late 1820s, much arable land had been left fallow. Even after the increased production brought on by commerce with the forty-niners, there was still land available for the military colonists. Initially, fallow land that was explicitly "irrigatable" was brought back into production as colonists negotiated the transfer of land from the Indians at San Xavier. When the vacant land ran out, the government began the process of expropriating land already under cultivation by current residents for the use of the new settlers. As might be imagined, howls of protest immediately went up, and rancorous litigation initiated by the long-time residents of the area disrupted enforcement of the new policy. The effectiveness of local government in Tucson deteriorated as controversy dominated the next two years.[3]

Mexican officials intended the new military colonies to further development in the region, which led quite understandably to a new effort to maximize the water supply in the Tucson Basin. In describing the Apache attack in December 1850, Captain Comadurán mentioned a large work party at a "dam" on the Santa Cruz River near the old San Agustín mission: "Approximately forty men of the town were at the dam and, lacking weapons, they sought refuge in the convent of the mission. Another ten, with their livestock, headed for San Xavier."[4] This cryptic reference to a dam on the Santa Cruz is left hanging. The type and specific purpose of the dam, and whether the dam was ever completed, remains in question, although it seems reasonable to assume from its location that damming the spring at the base of "A" Mountain, later the site of a dam and small lake, was the object of the construction.

Initially, the Sonoran government supported the military colonies more rhetorically than materially. Perhaps part of the reason for the neglect of the frontier had to do with the inexorable approach of the cholera epidemic. The disease approached with the same implacable certainty in the Mexican frontier as it had in the seaports of Europe. Cholera, caused by a bacteria that infects the lower intestine, results in severe diarrhea that may last a week, eventually killing the victim by acute dehydration. The bacteria spreads through tainted water and food—it thrives in wells too close to privies—and the most sure prevention is to establish a sanitary water supply. What made the illness so frightening in the mid-nineteenth century was that the bacteria was unknown, yet to be discovered by future scientists. Its effects were obvious, however, as it spread from seaport to seaport, from city to city. You could see it coming; the rising death tolls were unmistakable. The disease was reported in Tucson in July 1851, and added further disadvantage in the fight against the Indians beyond the relatively prosaic shortage of horses. The Tucson military colony lost nine soldiers to the disease, with others stricken but surviving only after a long convalescence. One of the casualties to the disease may have been the long-time commandante, Antonio Comadurán.[5]

Another reason for the effort to bolster the defense of the northern frontier was the increasing number of Americans seeking to reside in Sonora. By the summer of 1851 some of the sojourners seeking gold in California had come to realize that the fabled wealth was more dream than reality. Remembering they had passed obvious mineral deposits in northern Sonora on the way to California, some of the forty-niners headed back to the Mexican frontier to mine for gold and silver. Sonoran officials were quite concerned when, in July 1851, a group of over forty *Yanquis* set up residence several miles south of San Xavier for the purpose of mining.[6]

Mexican officials welcomed some Americans, however. In the fall of 1851, as cool autumn temperatures replaced the heat of summer, a group of Mormons settled at Tubac. The village had been recently reoccupied as a military colony following a particularly deadly Apache raid at Calabazas. The Mormons had been heading for California, but Mexican officials, seeking to develop the area, convinced them to stay in Tubac, their arguments bolstered by the season's pleasant climate and the promise of free

The artist Charles Shuchard, traveling with A. B. Gray's railroad survey in 1854, sketched Manuel María Gándara's hacienda and sheep ranch on the bluffs above the confluence of Sonoita Creek and the Santa Cruz River. This view from the south shows the San Caye-tano Mountains in the background. The row of cottonwood trees behind and to the left of the hacienda marks the course of Sonoita Creek before it joined the river out of the scene to the left. This area of "Calabaza" had been occupied by Indians for millennia and was one of the original Spanish visitas. It would become the site of the railroad town Calabasas in the 1880s, and today the community of Rio Rico spreads over the hillsides.

land. Although the Mormons did not settle permanently at the new mili-tary colony, Tubac had begun another cycle of growth, which was fur-thered when a group of peaceful Apaches from Tucson resettled in the region in 1853.[7]

In another effort to develop the valley and utilize the river, Manuel María Gándara, former and soon-to-be-again Governor of Sonora, estab-lished a sheep ranch at Calabazas with a specific commercial goal in mind.[8] Gándara sought to exploit the market for mutton that had developed in California. In 1852 he hired German immigrants in Hermosillo to work the ranch, and contracted for "5,000 sheep, 1,000 goats, 100 cows with calves, 100 brood mares, 10 yokes of oxen, 6 pack mules, [and] 10 horses for use." The ranch was in operation sometime in 1853. In the summer of 1854 Apaches attacked Gándara's ranch "killing fifty sheep and driving off the stock," but the ranch was still occupied by the German overseers in May 1855 when German sojourner Julius Froebel happened by on his tour of Mexico and the "Far West."[9]

In the not-so-distant past—twice, in fact—settlement at Calabazas and Tumacácori had increased, and downstream at Tubac water shortages, perhaps more perceived than real, had developed. By January 1854, rations

Charles Shuchard made three sketches of scenes in the Santa Cruz River valley, all of them more accurate in their rendering of the setting than those drawn by John Russell Bartlett, J. Ross Browne, or any other sojourner in the 1850s. In this view of "Tumaccacari," the riparian area along the river is clearly depicted behind the mission and its fields. In the background the Santa Rita Mountains are drawn fairly accurately, with the silhouette of Elephant Butte faintly visible at the far left of the range. The scene is remarkably similar to the view today of the old mission, given the renewed riparian area downstream from the international sewage treatment plant in Nogales.

for the troops at Tubac ran out, whether from real scarcity of water or a scarcity of labor due to a Sosa-type of laziness, we do not know. But in the face of the calamity the exasperated commander of the new military colony formed a supply train, utilizing mules and pack equipment from the Gán-dara hacienda at Calabazas, to fetch rations from Santa Cruz. As one might have expected, Apaches attacked the supply train near the old ranch of San Lazaro where the river makes its trademark U-turn. Only four of the seven soldiers dispatched on the mission returned to Tubac—with their lives but empty-handed. By the time a group of Texans passed through Tubac in September 1854 only a few soldiers and settlers remained. Although the colony somehow managed to survive the winter of 1854–55, Tubac was abandoned when American boundary surveyors passed through the mili-tary colony in the summer of 1855. Renewed settlement had survived a scant four years.[10]

The general state of development and commerce in the valley—and the subsequent demand on the river—remained slight. Estimates of Tucson's population varied from 300 to 600 residents through the early 1850s, while Tubac variously sprang to life and winked out. Tucsonenses maintained a meager 300 acres of agricultural production. Cultivation at Tubac was fitful. Intensive livestock grazing returned to Calabazas for a time, but Tumacácori remained only haphazardly cultivated. Agricultural production persisted at Santa Cruz at levels suitable for the residents' subsistence and the emigrant trade. Overall, demand on the river during this time fell far short of that required by previous levels of agriculture and commerce, and the returned river continued to flow at its most bountiful extent.

The next spate of Anglo observations came in the form of boundary commission surveys from 1851 to 1855, as well as surveys on behalf of proposed railroad construction. John Russell Bartlett toured the area as part of the U.S. Boundary Commission in 1851 and again in 1852 seeking to verify and mark the border established by the Treaty of Guadalupe Hidalgo. Once again the Anglo reports are problematic and inaccurate in places, but nonetheless provide an interesting contrast to official chronicles of the Mexican authorities. Bartlett mounted a veritable military campaign, with scores of wagons and herds of sheep and cattle in his train. A military contingent commanded by Lt. Col. Graham accompanied Bartlett, and Graham's report corroborates many of Bartlett's observations, and indicates the commissioner's imperious and dilatory nature as well.[11]

Bartlett's expedition first entered the area of the San Bernardino Ranch in September 1851, encountering the expected wild cattle. Bartlett had familiarized himself with the reports of the Mormon Battalion's passage through the area in 1846, and was aware that the soldiers and subsequent travelers had utilized the cattle herds to replenish their supplies. Bartlett's account supports earlier reports of diminished herds, the feral cattle population winnowed by the constant stream of traveler's through the region, as well as by hunts by Mexican settlers. Bartlett characterized the herds as "small, not more than six in each, led by a stately bull."[12]

From the San Bernardino Ranch, Bartlett's party entered the Santa Cruz River valley through the Canelo Hills and the San Rafael Valley. Heavy rain showers in that September, either from a lingering monsoon or a soaking El Niño event, hindered the caravan's progress toward the valley. On September 23, after two days without rain, the group crossed the low summit of the Canelo Hills, where the river emanated from springs on the southern slope. Viewing the broad valley beyond, Bartlett remarked on the thick grass and impressive stands of live oak. As later settlers noted, the terrain resembled the hill country of California, more or less perfect for raising cattle.[13]

After entering the San Rafael Valley through the Canelo Hills and traveling south toward the village of Santa Cruz, John Russell Bartlett remarked that "this valley was covered with the most luxuriant herbage, and thickly studded with live oaks; not like a forest, but rather resembling a cultivated park."

Corroborating many of the details of Bartlett's narrative, Lt. Col. Graham noted on September 21, "a real equinoctial rain, which lasted all this day." Upon entering the San Rafael Valley, Graham echoed the accounts of the beauty and fertility of the surroundings: "About 10 a.m., we reached the head of the beautiful valley in which Santa Cruz is situated. The grass was abundant and very luxuriant. The mules feasted on it as they traveled."[14] The accounts of thick grass verify the moist conditions that September, and also indicate that fewer cattle were grazing in the valley. As such, the stream was likely flowing most abundantly, unaffected by grazing livestock, feral or otherwise.

The Mexican officials at Santa Cruz provided Bartlett's group with 1,500 pounds of flour, but Bartlett remarked, "no other provisions could be obtained, so great was the dearth caused by the frequent incursions of the Apaches." The Indian raids had caused the Mexicans to abandon the village the year before, but residents had returned "a few months before our

The Santa Cruz River begins in the Canelo Hills and flows south through the San Rafael Valley. Bartlett may have encountered a scene such as this, the small stream meandering through grasslands toward the village of Santa Cruz. He noted that the town "possessed a fine valley and bottom land of the richest soil, and is irrigated by a small stream bearing its own name, which has its rise in springs about ten miles to the north, in the beautiful valley through which we entered the place."

arrival." Bartlett gave an enthusiastic description of the terrain, recognizing the great potential of the valley, with one major qualification:

It possessed a fine valley and bottom land of the richest soil, and is irrigated by a small stream bearing its own name, which has its rise in springs about ten miles to the north, in the beautiful valley through which we entered the place. It is admirably adapted for the raising of cattle and horses, as well as for all kinds of grain. Wheat, in particular, does remarkably well here. The *Chili Colorado,* of which such quantities are consumed in Mexico, grows here in perfection, and is said to be preferred on account of its superior piquancy to any raised in Sonora. . . . It is, however, a very sickly place, the inhabitants suffering from bilious fevers, in consequence of the proximity of a large marsh three miles west of the town. Many were ill at the time of my visit, and I was desirous to get away as soon as possible.[15]

Graham once again repeated some of Bartlett's observations, adding his own views of the "squalid" conditions of the settlement. Beneath his derogatory characterizations, however, Graham unwittingly recorded the clear presence in Santa Cruz of shrewd and experienced traders. Regardless of the appearance of the residents, destitute or not, they were holding out for high prices for food supplies, obviously recognizing that the commercial realities indicated a seller's market. For example, Graham first reported that "a few pigs and some poultry were seen about, but their owners seemed unwilling to sell them." In their recurring sweeps through the town in search of supplies, Graham reported very limited success. After one encounter, Graham indignantly noted, "A half-grown hog was offered to us for the enormous price of eight dollars, and a common-sized hog at twelve dollars, which we declined giving." Graham was correct to note the inflationary prices. A few years previously forty-niners were complaining about paying five dollars for a puny "grunter." The next day Graham succeeded in purchasing eighty pounds of dried beef, sufficient for the next leg of his detachment's journey, but the price of thirty-seven and one-half cents per pound for the "very bad dried beef" rankled. Clearly, supplies could be had in Santa Cruz, but after more than six years of trading with Anglo visitors, the residents of Santa Cruz were driving hard bargains. In a final disgruntled reproach for the residents of the military colony, Graham criticized the "abominable howling of the numerous dogs at night," which "annoyed us very much." No doubt the Americans left Santa Cruz with a great sigh of relief, and the Santa Cruz residents shrugged at their passing.[16]

Lurking "bilious fever" and Santa Cruz's limited supplies caused Bartlett to continue on to Magdalena, seventy-five miles to the south, in search of more flour and sugar. Graham and his military detachment returned to the copper mines in New Mexico, then traveled on to El Paso. From Magdalena, Bartlett eventually traveled to Guaymas, boarding a ship that took him to California, where he sojourned for the rest of the year. The next summer, in July 1852, Bartlett was back in Arizona, crossing the Colorado River at Fort Yuma, and heading up the Gila River to the Pima Indian settlements near the junction with the Santa Cruz River. After a stay of ten days in the Maricopa villages and five days with the Pima Indians, on the afternoon of July 13, Bartlett's party left the Gila River for Tucson.

On his second entry into the valley, Bartlett followed the probable path of the first Paleo-Indians and later fur trappers. Although much changed from the terrain of the late Pleistocene, Bartlett traversed a landscape not much different from the one observed by the trappers of the 1820s, outside the variations of normal seasonal change. The trappers had entered Tucson in the winter, Bartlett traveled in mid-July, just as the summer monsoon rumbled to life. In an example of our sometimes startling continuity with distant observations, Bartlett described a thunderstorm near Picacho Peak in terms any motorist today heading east on Interstate 10 would recognize:

> As the sun sank below the horizon, the dark cloudbank which we had observed far to the south ascended, and we could see the rain already falling on the distant mountain. Night now set in; the thick clouds rose higher and higher, and before nine o'clock had completely obscured every star. . . . But nature's light-house opened its portals, and the vivid lightning flashed around us. . . . Peals of the most terrific thunder burst upon us, leaving scarcely an interval of repose. Next came violent gusts of wind, accompanied by clouds of sand and dust, reminding one of the African *simoom*. The wind was from the south, and brought the sand directly in our faces. To avoid it was impossible. . . . Lastly came torrents of rain, and this terrific storm was at its height.[17]

The summer rain hampered steady progress of the heavy wagons in the caravan, but the plentiful rain, continuing the next night, made a quick sprint from the Gila to Tucson unnecessary. Puddles and pockets of water occurred regularly along the road.

Near Picacho, Bartlett's party met a group of emigrants from Missouri heading for California. This encounter presaged a common occurrence during Bartlett's journey up the valley: the presence of wagon trains and groups of travelers regularly spaced along the road, traveling in the opposite direction. Near Tucson Bartlett encountered the group led by "Mr. Coons, an American, on his way to San Francisco with 14,000 sheep." Coons employed an armed party of sixty drovers, including forty-five Americans and fifteen Mexicans. This herd of sheep and its attendants placed a transient demand on grass and water in the valley, in addition to the demand created by all the travelers, with their own herds of livestock and draft animals. Despite this demand on the resources of the valley, Bartlett reported plentiful grass along the road.[18]

Bartlett described Tucson as a besieged community of about 300 residents barely eking out an existence in the face of constant Apache raids. He called the valley "fertile" and the lands "rich." He remarked the valley was once "extensively cultivated; but the encroachments of the Apaches compelled the people to abandon their ranchos and seek safety within the town. The miserable population, confined to such narrow limits, barely gains a subsistence." Bartlett made explicit reference to the "copious supply of water" from the springs at the base of Sentinel Peak. He misidentified the abandoned village of El Pueblito and the old Convento as an abandoned hacienda. Given the resources available, the produce of the valley was potentially plentiful: "Wheat, maize, peas, beans, and lentils are raised in perfection; while among the fruits may be named, apples, pears, peaches, and grapes. The only vegetables we saw were onions, pumpkins, and beans; but in such a fertile valley all kinds will of course do well." Bartlett sketched the valley so as to catch the "agreeable landscape." He wrote in his journal, "The bottom lands are here about a mile in width. Through them run irrigating canals in every direction, the lines of which are marked by rows of cotton-woods and willows."[19]

On July 19 Bartlett left Tucson and traveled to San Xavier, which he described as a "truly miserable place," inhabited by Papago Indians in "eighty to one hundred huts, or wigwams, made of mud or straw." Bartlett remarked that the fertile valley, so clearly cultivated in the near past, was now largely fallow.[20]

As the group continued their southerly journey, more rain hampered their progress. The road to Tubac became muddy, and the heavy wagons slogged along. Bartlett remarked that the river channel was dry despite all the rain, and he considered this an indication of "the uncertainty of the streams in this country." Clearly Bartlett did not understand the geology of the basin in which the river traditionally sank just north of Tubac, not to return until the rising bedrock near Martinez Hill created the marsh that gave rise to the streams through the Tucson Basin. In his pronouncement about the "uncertain" streams, Bartlett firmly placed himself within the premodern vision of the river, looking no further than the surface flow. Later surveyors would hint at a changing perception, but Bartlett gave no sign of understanding the groundwater hydrology beneath the valley. In his view, the bountiful rain had simply saturated the surface of the desert, and the group enjoyed the resulting pools in low spots along the river's course.[21]

At Tubac, Bartlett noted the poor and dilapidated condition of the presidio, reoccupied during the preceding year. The commissioner considered the agricultural potential of the Tubac area to be similar to that of Tucson, given the river "full of water," a potential going to waste due to the depredations of the Apaches. But Bartlett also attributed the poor condition of Tubac to be the result of unreliable water resources. Bartlett recounted the experience of the Mormon settlers at Tubac as a cautionary tale about the fitfulness of rain and river water in the semiarid environment. The Mormons had been convinced to stay in Tubac by the commandante, who offered them rich fields already serviced by acequias. As Bartlett recounted, the commandante assured the Mormons that "they would find a ready market for all the corn, wheat and vegetables they could raise, from the troops and from passing emigrants." The Mormons went right to work, ploughing and planting the lands, but "the spring and summer came without rain: the river dried up; their fields could not be irrigated; and their labor, time and money were lost."

Assuming it unlikely that laziness doomed their effort, what was the cause of the Mormons' failure to revive agriculture in Tubac? Could the culprit have been the ineffectiveness of Spanish and Mexican acequias, or could the river simply have varied so much from one season to the next? Granted, rainfall amounts vary significantly from place to place in the region, but given the accounts by Bartlett and Graham of a bountiful monsoon the previous September, and more recently, Bartlett's report of the river "full of water" in the vicinity of Tubac that July, it is safe to assume a sufficient rainfall amount for that year even if winter precipitation was somewhat below normal.[22]

Thus either Bartlett or the Mormons had exaggerated or profoundly misunderstood their predicament. If the Mormons had missed any rain during their brief stay along the Santa Cruz River, it must have been during the very early spring of 1852—a normally dry time in the valley—not during the "spring and summer," as Bartlett had reported. Bartlett had encountered the Mormons and heard their tale in California, near Santa Isabel, in late May, during his journey from San Diego to Fort Yuma. The settlers were on their way to join the Mormon community at San Bernardino. To have arrived at Santa Isabel in May, the Mormon settlers from Tubac must have left the Santa Cruz River valley no later than mid-April, and so had not persisted long enough to experience the summer monsoon. In this the farmers probably mirrored Bartlett's assumptions about the region, expecting March showers to water their crops, ignorant of the weather patterns in the region that dictated a typically dry spring. As Bartlett's account made clear, by July 1852 another monsoon was thundering through the valley.[23]

Bartlett learned from a group of Arkansas emigrants traveling north along the river that the environs to the south had benefitted from abundant rain, and that plenty of grass was available along the road. In the area of Tumacácori and Calabazas Bartlett described the valley as particularly lovely: "The bottom-lands resembled meadows, being covered with luxuriant grass, and but few trees. The immediate banks of the river, which is here as diminutive as near Tucson, are lined with cotton-wood trees of gigantic size, resembling our largest elms. In some places there are large groves of these trees, rendering this part of the valley the most picturesque and beautiful we had seen."[24]

Once again Bartlett explained, erroneously, that the abandonment of Calabazas and Tumacácori had resulted from limited water supplies. He noted that the earlier productivity of the area had deprived both Tubac and Tucson of scarce water, assuming that the river naturally flowed continuously from Tubac to Tucson. Bartlett did not understand that the river rarely flowed continuously to Tucson, and that cultivation in the upper basin had little effect on agriculture in the middle basin. In his explanation of the river's fitfulness, Bartlett displayed both his premodern perception and his northern and eastern bias. "This is the difficulty with these small water-courses; for having few or no tributaries to keep up the supply, as our northern streams have, and frequently running a course of several hundred miles before they terminate, their water cannot be drawn off without destroying the crops below them, and even depriving the people and animals of water to drink."[25] Although Bartlett's statement correctly notes the problem with a legal doctrine concerning water rights based on prior appropriation in a semiarid environment, he was erroneous in his particular observations of the river.

South of Calabazas, the group encountered two more wagon trains from Arkansas heading for California. Each party was composed of twelve to fifteen wagons pulled by oxen. Bartlett again remarked on the beauty of the valley, tempered by an awareness, if only partially correct, of the temporary nature of the scene:

> The country grew more picturesque and diversified, exhibiting alternate valleys and gentle hills. In the former were groups of large walnut-trees, whose deep green foliage presented a striking contrast with the lighter and yellowish hue of the cotton-woods, and the brighter green of the willows. The intermediate spaces between the hills and the stream exhibited a luxuriant growth of grass. But it must be remembered that the enchanting aspect which every thing now wore in this valley does not continue. It was the rainy season, when vegetation presents its most attractive garb. In a few weeks the daily showers would cease, and the parching sun would dry up every thing but the cacti, which do not seem to be benefitted by rain, and the large trees of the valley, which find moisture enough in the earth to sustain them until the rainy season again comes round. The grass then withers and dies, and the stream furnishes barely water enough to supply the immediate wants of the people.[26]

Another aspect of Bartlett's limited understanding of the region and its flora concerned his view of the native grasses. Certainly much of the grass dried during the months of slackened rainfall, but varieties of perennial grass continued to provide sustenance for livestock and animal species for months after turning brown.[27]

Before Bartlett's party reached San Lázaro they encountered another group of emigrants in a train of nine wagons, each pulled by ten oxen. The settlers understood that ten oxen were more than necessary to pull the wagons, but they told Bartlett that they were taking the oxen to market and it was easier to harness them to the wagon than drive them separately. Another large group from Arkansas was camped at San Lázaro. Once again Bartlett remarked on the "extensive orchards and fertile grounds" near the abandoned hacienda. From San Lázaro, the commissioner's party traveled nine miles near the river "over a rough road" into Santa Cruz.[28]

Santa Cruz remained in the pinched circumstances Bartlett remembered from the previous September. Indians skulking about had caused the residents to stay close to the town, which caused the commissioner to assume that agricultural production was limited to "barely enough for their own subsistence." Nonetheless, Bartlett reported that flour was available for sale in the town, and at one point the village's priest paid a kindly visit to Bartlett, bestowing gifts of a bottle of wine and "another of excellent vinegar . . . together with a few vegetables." Bartlett's group stayed four days in Santa Cruz. Another group of Arkansas emigrants—twenty wagons, each drawn by ten oxen—passed through town during Bartlett's stay. These Americans found no trade goods available in Santa Cruz, probably due more to the American's ethnocentrism than any shortage in local production. Early in their visit, the emigrants had offended the local priest by playing cards on the church grounds and behaving with a "boisterous manner and insolence." As the Mexican residents in Santa Cruz had shown in the past, if they found American customers unseemly or disrespectful, trade goods disappeared, and blank looks of helpless poverty greeted the visitors.

Bartlett's party left Santa Cruz for the presidio at Janos on July 28, 1852. The journey from Santa Cruz to Janos was notable for two circumstances reported by the boundary commissioner. First, frequent rain continued to hamper their travel, and second, Bartlett reported areas formerly noted for their large herds of wild cattle were now devoid of any livestock or any settlement. Bartlett expressed his own disparaging view of Mexican culture in this observation, correctly attributing the lack of settlement to Apache depredations, but failing to recognize the combination of factors that had given rise to the wild herds, and then caused their slow demise. In his mind, simply, the lack of development would be reversed and the return of large herds of grazing livestock would be accomplished, once the area was placed "in other hands."[29]

Next into the area were the parties of Americans conducting surveys in conjunction with proposed railroad routes and the new boundary established by the Gadsden Purchase. The formal name for the purchase is the Treaty of Mesilla, referring to the agreement by President Santa Anna to sell 30,000 square miles of northern Mexico to the United States for

$10 million in 1853.[30] A party led by Lt. John G. Parke began surveying railroad routes in January 1854, and were camped near Tucson in February. During the jornada from the Gila River—the Army surveyors had entered the region from San Diego and Fort Yuma—Parke reported finding "pools of rain-water" near Picacho Peak, and that a "smart shower" awakened the group on their second day along the road. The group entered Tucson on February 20th, camping about two miles south of the town, "on the bank of a clear running brook, with [an] abundance of grass and wood." The surveyors procured "a good feeding of corn" from the settlement, which Parke estimated had a population of about six hundred. Parke characterized the production of the valley as restricted due to Apache raids and the limited water supply. "They raise chiefly corn and wheat, cultivating about three hundred acres of rich soil by irrigation from a stream which has its source near the mission of San Javier [sic], 8.5 miles to the south; and although it flows past our camp with a depth of one foot, and width of six feet, its waters nevertheless disappear a short distance below the town, either consumed in irrigation or absorbed by the sands." The engineer even recorded the temperature of the river water on a brisk, thirty-two-degree February morning as being sixty-two degrees.[31]

There are hints of a modern perception of the river and its aquifer in Parke's report. When the engineer stated that water in the channel either served irrigation or was "absorbed by the sands," he may have been expressing a tacit understanding of the nature of the aquifer. The surveyor correctly identified the source of the river through the basin, although it seems curious that he talks of a single stream through the valley rather than a split stream (later identified as the main branch and spring branch). Likely, Parke considered the divisions in the stream to be dictated by irrigation demands. Without placing too much significance on this surveyor's observations, it is interesting to note that someone in 1854 was sticking a thermometer into the river, indicating by that act a heightened technological presence in the valley, and perhaps even an increased curiosity about the river and its underground circumstances. On February 22, the surveyors left their river campsite and headed east along Cooke's trail to the San Pedro Valley.[32]

Another railroad survey party led by Colonel Andrew B. Gray was in the vicinity of Tubac in April 1854, but Gray's journal contains few references to the physical settings of the river valley, except for elevations and prospective grade estimates. One exception is Gray's offhand reference to Santa Cruz, which he remembered from an earlier visit as a member of the Bartlett expedition: "The atmosphere is pure and healthy, and the cli-

In 1854, Charles Shuchard depicted a scene of lush vegetation at his Tubac campsite, which was indicative of a flowing river and a saturated aquifer. Here and in his drawing of Tumacácori, the settings resemble views of the renewed riparian area today, which is as abundant as the one Shuchard drew almost 150 years ago.

mate agreeable winter and summer, except in the immediate vicinity of the town of Santa Cruz, where there are swamps hemmed in by high mountains."[33]

The most vivid descriptions of the Santa Cruz River valley in Gray's report appear in the series of sketches made by the party's artist, Charles Shuchard. Shuchard's sketch "Camp View Near Tubac, Sonora" conveys a sense of luxuriant vegetation along the river. Impressive and stately trees tower over the campfire, which is set among a clearing surrounded by cattails and underbrush. Shuchard also sketched the Gándara hacienda at Calabazas and the mission at Tumacácori, showing flourishing cotton-woods along the river's course by the sheep ranch and the church.[34] Perhaps ironically, the renewed riparian area in the vicinity of Tubac (to be discussed in later chapters) today surpasses Shuchard's drawings for its lush vegetation and opulent stream flow. At the very least, visitors to that stretch of the river today can easily imagine the fertile scene that welcomed the railroad surveyors and their accompanying artist.

In the last months of Mexican rule prior to the transfer of control to the U.S. government, the same patterns continued on the frontier. Apache raids plagued the settlers and military colonies, while a continuing stream of Anglo travelers passing through the valley encouraged increasing development and commerce. A group from Texas driving a herd of cattle to California passed through the valley in September 1854. James G. Bell kept a journal of the trip, making what were, by now, familiar racist references to the Mexican residents. As to the circumstances of the valley, Bell noted that one night in mid-September was "foggy and cool," perhaps indicating a continuing wet cycle in the valley.[35]

The cattle drive passed the Gándara sheep ranch on September 13, and Bell noted that Apaches had made a raid on the place, killing fifty sheep, "a few weeks since." At Tubac a few soldiers remained on hand, but soon after the cattle train passed through Tubac, the Mexican authorities ordered the military colony's soldiers to Santa Cruz, in preparation for the transfer of sovereignty to the Americans under the Gadsden Purchase agreement. Officer suggested that a few settlers, both Hispanic and peaceful Apache, may have remained in Tubac for a short time after the soldiers left, but by the summer of 1855 the village was abandoned.[36]

The cattle drive was camped south of Tucson on September 20, 1854. Bell apparently had encountered a group of American miners in the area, who told the Texans of "an abundance of gold" nearby. Tucson residents, Bell reported, offered "figgs [sic], apples, quinces, pomegrantes, [sic]" as well as "other fruits." In general, however, Bell was quite disdainful of the town and its inhabitants, expressing a clear ethnocentrism. On September 21, the drovers moved their herd ten miles north of the town to a place where they found sufficient water in "pockets" near the road, and continued on to the Gila River settlements in the following days.[37]

A new boundary commission under the overall command of Major William H. Emory traveled through Tucson on their way to Nogales in 1855 to complete surveys of the new border under the Gadsden Purchase. Lt. Michler led a detachment of surveyors under Emory's command. Emory's 1856 report contains general references to the terrain and environment of the Santa Cruz River valley, while Lt. Michler's report contains specific references to the valley's conditions and circumstances.

Emory had traveled extensively in the Southwest, including an 1846 journey with Kearney's Army of the West, prior to his appointment as boundary commissioner to replace John Russell Bartlett in 1854. His previous impressions informed his description of the territory acquired in the new treaty:

It is very possible the whole of the new territory, except the region of desert country referred to . . . may be brought under the influence of artesian wells and made productive; but until that is the case, agriculture must be confined to the beds of the river, where the land is below the water-level. There are many tracts of this kind of surpassing richness, but of limited extent, on the Rio Bravo, on the Rio Gila, on the San Pedro, and on the Santa Cruz. Those which are most conspicuous, and which are at present in a very advanced state of cultivation, are the Mesilla Valley on the Rio Bravo, the Valleys of Tucson and Tomacacori [sic] on the Santa Cruz, and the settlement of the Pimos [sic] on the Gila river.[38]

Emory had ordered Lt. Michler to survey that portion of the new border in the vicinity of the Santa Cruz River. Michler and his small detachment traveled to Arizona by way of San Diego, Fort Yuma, and the Pima villages on the Gila River, finally arriving at Tucson in June 1855. Along the way from the Gila Michler remarked that *tinajas* (water pockets) near Picacho Peak remained "filled with water for short periods after the rainy season." He failed to mention if they found water in the pockets, but it is safe to assume they did not, given that Michler was of the opinion that the rainy season had not yet started.[39]

Michler's description of Tucson conformed to earlier accounts describing it an oasis of agricultural production. Significantly for Michler, the production in the valley had apparently been accomplished prior to the start of the summer monsoon. Viewing the dry and barren surroundings of the lower basin, Michler may have mistaken this and the drier central valley of the Colorado and Gila as indicative of the entire region, as became evident in his description of the middle and upper basins of the Santa Cruz River, "clothed with rich green verdure," in comparison to the "bleached barrenness of the Colorado and Gila." The jornada from the Gila had left the Americans "famished," and made the approach to Tucson memorable:

Several miles before reaching Tucson you strike the bed of the Santa Cruz river, but the stream is subterraneous until you reach the town. The latter is inhabited by a few Mexican troops and their families, together with some tame Apache Indians. It is very prettily situated in a fine fertile valley at the base of the Sierra de Santa Catarina [sic]. Some fine fields of wheat and corn were ready for the sickle. Many varieties of fruit and all kinds of vegetables were also to be had, upon which we indulged our long-famished appetites.[40]

The Americans stayed in Tucson the entire month of June, fattening their mules and replenishing their supplies. During the crew's hiatus, Michler left the group to meet Major Emory and the Mexican commissioner near Los Nogales, sixty-nine miles south of Tucson. Traveling up the valley, Michler offered brief descriptions of San Xavier and Tubac, the latter of which he found "a deserted village." At Tumacácori, Michler found the abandoned church, "another fine structure," standing "in the midst of rich fields; but fear prevents its habitation, save by two or three Germans, who have wandered from their distant fatherland to this out of the way country." These Germans may have been part of the Gándara sheep ranch personnel, but Michler made no reference to the Germans as attached to the "Rancho de las Calabassas [sic]." Traveling up "the pretty little valley of Los Nogales," Michler found the U.S. Commissioner's camp. From Michler's references to the lush vegetation, an early monsoon would seem to be evident. Michler is explicit, however, that the monsoon had not yet commenced in full force, merely giving "indications" of its "commencement." On June 26, Michler and his Mexican counterpart, Mr. Jiménez, began surveying the next section of the boundary line, which could only be undertaken during the rainy season. The monsoon had arrived more-or-less on time, according to Michler, in late June. "At any other period of the year it would have been impossible to attempt this section of the work, as there is little or no permanent water in the neighborhood of or along the whole length of this line of two hundred and thirty-seven miles." Michler and Jiménez were the surveyors who plotted the line on the map that intersected the Santa Cruz River twice.[41]

In the last days of Mexican control, American sojourners passed through the Santa Cruz River valley, and other Americans arrived with the notion to stay. Several bought plots of land in Tucson in transactions "certified" by the Mexican commander. As Officer reported, plots were selling for ten to twenty pesos. Most of the Hispanic residents of Tucson welcomed the Americans' arrival. To the beleaguered settlers, the transition to U.S. sovereignty brought the prospect of protection from the Apaches, the possibility of broadened trade and commerce, and hopeful relief from the fractious land-tenure disputes brought on by the most recent Mexican regime. But not all the residents of Tucson welcomed the transition. The last Mexican troops left Tucson in March 1856 heading south to join the former Tubac soldiers now stationed at Santa Cruz. Settlers unwilling to stay in the area also joined the migration south. While excited Tucsonences commenced a fiesta as American soldiers raised the United States flag at the old presidio, Mexican soldiers and settlers determined to remain loyal to Mexico marched south into an unusually heavy March rainstorm.[42]

The area's transition to United States control presaged a change in the river's circumstances, even if the stream itself voiced no opinion on the matter of sovereignty. As mining, ranching, and farming activity in the valley increased, the decades of slackened demand on the river came to an end. Some of the new settlers brought with them a new, modern perception of the river and possessed an eye toward tapping the aquifer to the fullest extent possible. The Santa Cruz River's aboveground days were numbered.

7

New Residents, Old Problems
1856–1880

As increasing numbers of Anglo merchants, miners, and ranchers entered the valley in the late 1850s with the intent to stay, a new wave of development and commerce occurred along the banks of the river. The ebb tide of settlement characterized by the abandonment of Tubac in 1855 turned to a flood tide of development spurred by renewed interest in ranching, mining, and agriculture. Miners, especially, were more likely to bring a new perception of the valley's water resources, aware from the occurrence of wet mines that a huge reservoir of groundwater underlay the fitful surface flow of the river.

Anglos enjoyed relative peace with the Apaches early on, but as the number of American inhabitants slowly increased, Apache animosity previously directed toward Hispanic settlements came to include raids and attacks on Anglo ranches, mines, and farms. The U.S. military increased its presence in the area and offered settlers some modicum of protection against the Apaches, but the advent of the Civil War in 1861 caused retrenchment. The raids led by Cochise and other Apache leaders increased, and development in the valley receded as many recent arrivals abandoned their mines and other entrepreneurial endeavors. As a result, the river flowed in its ancient pattern through the 1870s in the upper basin, the meager human presence utilizing hardly a trace of the water in the channel. In the middle basin, on the other hand, settlement increased despite the Apache raids—primarily in Tucson—and the years of slackened demand came to an end. Settlers constructed dams in the stream channel to power mills with water flowing through the millraces. By 1879 the source of domestic water supply in Tucson, soon to be nicknamed the Old Pueblo, became the subject of debate among engineers, capitalists, and politicians. As always, the river remained at the center of life in the valley, in circumstances that varied over its course and through time.

Among the new immigrants were settlers who would profoundly influence the river's circumstances. Solomon Warner and William Rowlett arrived in Tucson in 1856 eleven days after the Mexican troops left. Warner was a merchant and developer, and immediately began a business partnership with Mark Aldrich, who was operating Tucson's first mercantile store. Entrepreneurs William Rowlett and his brother Alfred first built a "low earthen dam" that gathered water from several springs feeding a cienega south of Sentinel Peak, creating a lake, later known as Silver Lake. The next year they built a flour mill to the west of the lake. The brothers had connected three ditches emanating from the cienega into one stream, which they used to turn the water wheel of the mill. The water discharged from the mill then flowed into Silver Lake, and from there into the public acequia that serviced the fields near Tucson. The reservoir allowed the Rowletts to operate the mill independently of the irrigation needs of the farmers west of town. Solomon Warner would later construct another lake and mill in Tucson that impounded irrigation water and eventually figured prominently in litigation over water rights in the 1880s.[1]

Both of these efforts to harness the surface flow of water in the basin constitute further examples of the archaic perception of the river. Although Warner would later make references to the underground aquifer, it was not in relation to his own damming endeavors. The Rowletts' left no testament to their perception of the river, but from the extent of their efforts, their vision of the river's resources went no further than the surface flow.

Another early immigrant, Sam Hughes, a native of Wales, arrived in Tucson in March 1858. Hughes became a merchant and developer, but did not initially engage in direct manipulation of the river, such as Warner and the Rowletts had. From his later efforts in riverine manipulation, however, Hughes showed he too held a premodern vision of the river. A long-time resident of Tucson, Hughes left accounts of life in the village from the beginning of American sovereignty. In one instance, Hughes described the winter of 1858–1859 in Tucson as very wet and cold. "The waters of the Santa Cruz were so deep that a flat boat could be navigated probably clear to the Gila at Maricopa," observing as well that "the Rillito was a mile wide." During floods the river has since surged to such an extent, but Hughes's account may have been exaggerated, since he also reported to have met Mose Carson (Kit's brother), who Hughes claimed "had spent that last winter [1857–1858], trapping beaver on the Santa Cruz and Rillito Rivers," even though there is no evidence of the presence of beaver in the

Santa Cruz Valley even back to the days of the Hohokam. Hughes made these reminiscences late in life, and so might be forgiven for certain flights in memory. On the other hand, this reference to beaver could have been attributed to Carson, with the somewhat gullible Hughes taken in by the frontiersman's braggadocio.[2]

The issue of the presence of beaver in the valley is further confused by a brief reference by G.E.P. Smith in 1910. Smith mentioned beaver in the "historical sketch" of his study of groundwater and irrigation in the Rillito Valley: "The river course was indefinite,—a continuous grove of tall cottonwood, ash, willow and walnut trees with underbrush and sacaton and galleta grass, and it was further obstructed by beaver dams." Unfortunately, Smith made no citation of sources for his historical sketch. The reference to beaver dams occurred in his discussion of the conditions of the valley when the first settler, "an Arkansas pioneer," began cultivating a small parcel along the Rillito in 1858. Smith was writing in 1910, fifty-two years since the Arkansas pioneer had plowed his first farrow, and a decade after beginning his career at the University of Arizona in 1900. Likely his reference to beaver was based on reminiscence or hearsay rather than actual observation or documented source.[3]

Two years after the Arkansas pioneer began farming along the Rillito, in 1860, William S. Grant, an investor and entrepreneur from Maine, acquired the Rowlett brothers' mill, which by then was supplying flour to the military garrisons in the area. To keep pace with the military's demand for flour, Grant reconstructed the dam and mill, and built a second mill at Silver Lake just before the Civil War. In 1861, retreating Union forces following a scorched-earth policy destroyed Grant's mills to prevent them from falling into Confederate hands. Then in a complete turn around, the returning Union forces in 1862 repaired one of Grant's mills in order to keep the troops supplied with flour.[4]

In a happy circumstance for historians of Tucson and the Santa Cruz River, the Union commander, Col. Joseph R. West, ordered a map to be drawn to record properties subject to requisition or confiscation. Major David Fergusson got the job of map making, and his maps and report offer a benchmark characterization of the river status in the Tucson Basin in 1862. Approaching Tucson, the river disappeared into a system of irrigation canals, the main ditches running roughly south to north, labeled in his map "La Acequia Madre Primera," "Acequia Madre Segunda," and "Acequia Madre Tercia." La Acequia Madre Primera was the ditch closest to town with the others ranging to the west. The fields in the floodplain encom-

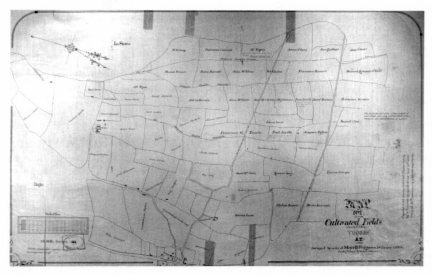

Major David Fergusson's 1862 maps of Tucson are the first surveyed maps drawn after Tucson became part of the United States. His map of the cultivated fields west of Tucson offers a glimpse of the old irrigation system. The river disappeared into three main irrigation ditches, with the *primera* ditch the one closest to town. The 800 acres under cultivation extended to St. Mary's Road, with fields north of the road farmed only "when there is an abundance of water."

passed over 800 acres. Fergusson's maps also indicate that the fields north of present St. Mary's Road in central Tucson "are only cultivated when there is an abundance of water."[5]

Fergusson also surveyed the route from Tucson south into Lobos Bay in Mexico, because the commander of the Union forces in Arizona was seeking to convince the War Department that supplying the troops in Arizona would be easier by way of the Gulf of California, rather than overland from Kansas. In his report to the War Department, Fergusson described the river south of the San Xavier mission through Sahuarito [*sic*] to Roade's Ranch (near Canoa). First Fergusson noted "El Rancho Viejo" 1.71 miles south of the mission: "Good road through meadow; running water 200 yards to left of road; wood; grass; an abandoned ranch belonging to San Xavier mission." Next came "Punto de Agua, also known as Struby's rancho." 0.79 miles south of El Rancho Viejo: "Good road through dense mesquite; running water 150 yards to left of road; wood and grass; shade; ranch abandoned." Sahuarito, "also known as Columbus," was 8.38 miles beyond Punto de Agua: "a deserted rancho; good road, somewhat dusty; grass; an old well caved in, water near the surface; wood." Roade's Ranch was

8.56 miles south of Sahuarito: "Good road; at the forks of the road take the left; permanent water in well, about 12 feet deep; a good curb well should be made here to *secure* [Fergusson's italics] water for large trains; grass, cotton-wood; shade. Ranch abandoned." Betancourt took the major's reference to running water at El Rancho Viejo and Punto de Agua to indicate the so-called "stream branch" in the east, and the "mainstem" on the west side of the floodplain. Whereas this may be correct, there is no clear reference in Fergusson's report that the streams were independent and separate. Fergusson may have been describing a continuous stream in its relation to the wagon road, consistently to the left, or east, of the road.[6]

Other immigrants into the valley left accounts of their travels in the tradition of the earlier argonaut reports: romantic, ethnocentric, marginally reliable, and generally laudatory of the terrain and the river's potential. Access to the region gradually increased with the commencement of stage service in 1858, although travel remained arduous and challenging, especially to eastern sensibilities. Three Americans, Charles Poston, Phocion Way, and J. Ross Browne, traveled through the valley from 1856 to 1864, keeping journals and writing accounts of their travels. Little differentiates these reports from the earlier observations by forty-niners and government surveyors, and so only a brief overview is necessary.

Charles Poston established the headquarters for his mining company at the abandoned presidio of Tubac in 1856. A new spurt of occupation at Tubac ensued, even though little actual mining took place in the area. Poston was a first-class promoter and salesman for the territory, but much less successful in furthering actual development. A talent for hyperbole if not literary flair appeared in his description of the river at Tubac, calling it "a stream as large and as beautiful as the Arno." Poston continued the classical allusion as his entourage entered the deserted village, "It was like entering the ruins of Pompeii."[7] No doubt exaggerating, but perhaps indicative of increasing activity in the valley, Poston wrote, "By Christmas, 1856, an informal census showed the presence of fully a thousand souls . . . in the vicinity of Tubac."[8] Passing through Tubac in 1858, Phocion Way reported "about 150 inhabitants, three-fourths of whom are Mexican," which brings Poston's hyperbole somewhat back to earth.[9]

The presence of renewed settlement in Tubac placed only slight demands on the river. In the spring of 1857, the headquarters staff planted a two-acre garden watered by a canal from the river. Perhaps Poston had recruited one of the German overseers from the Gándara Rancho to supervise the small truck farm: "By the industry of a German gardener with two Mexican assistants, we soon produced all vegetables, melons, etc., that we required, and many a weary traveler remembers, or ought to remember, the hospitalities of Tubac."[10]

In 1857 the first regularly scheduled mail service began. The San Antonio and San Diego Line never prospered, but was followed in 1858 by the Butterfield Line, which established itself as a prominent fixture in Tucson and territorial Arizona. The stage and mail lines brought sojourners and new residents. Phocion Way traveled through Tucson in 1858 as a passenger on one of the first regularly scheduled passenger and mail runs. Way left Cincinnati in May and arrived in Tucson about a month later. He had traveled to Arizona as an agent and employee of the Santa Rita Mining Company, which was an "outgrowth" of Poston's Sonora Mining Company. Way stayed in the territory only two years before returning to Ohio disenchanted with the prospects of quick riches in the Arizona silver mines.[11]

Way's narrative conformed to the style of the times, occasionally exaggerated and generally critical of Mexican culture and society. His effort at literary flair also intruded; for instance, in describing the scene as the stage approached the San Pedro Valley. The particularly rough traverse through the mountains reminded Way "of Bonaparte crossing the Alps."[12]

Way arrived in Tucson on June 12, 1858, and had nothing good to say about the town. His negative perception was partly determined by the premonsoon heat in the basin. "There is very little air stirring, and if hell is any hotter than this I don't want to go there."[13] No hotel or tavern graced the place, and there was "not much to eat in the d——d town." In a typical display of shortages, not quite absolute, the American reported "no fresh meat to be had," and then later expressed great joy and amazement when the "captain" of the mail run obtained a "roast of beef." Tucson residents, akin to the denizens of Santa Cruz, understood the commercial realities of a seller's market.[14]

To Way, the river through Tucson that June was barely a "creek." He complained that the water was "alkaline and warm." He also made a passing reference to well water in Tucson. Traditionally, Tucsonans acquired their domestic water from El Ojito, an artesian well on the eastside of the floodplain along the road to the San Agustín mission. Wells had proliferated inside and outside the old presidio walls prior to the Civil War, and Way made reference to a small group of Anglos who "have dug a well and procure tolerably good water, which they use." Many wells flowed copiously at first, then went dry or became tainted with alkali, causing Tucsonans much consternation. Besides the aggravation caused by scarce or unpalatable water, the number of dry holes in town, visible during daylight hours, proved troublesome to nighttime pedestrians. By the 1870s, a hand-dug well on South Main Street provided the water which Adam Sanders and Joseph Phy sold at five cents a bucket. The two entrepreneurs filled an iron tank on a wagon from their well and traveled daily through town selling water. Within twenty-five years municipal water use in Tucson would progress from well water sold by the bucket, to a piped supply tapping the aquifer. When the mains were first opened in September 1882, an almost immediate decline in the water table downstream resulted, but this was a circumstance unforeseen by Way, or by anyone else for that matter, in 1858.[15]

Way's mining activity took place in the Santa Rita Mountains, and he much preferred the environs of Tubac and the cooler elevations of the mountains. Ethnocentrism and a tendency to gripe about most anything dominates the remainder of Way's journal. He never did feel comfortable in the Southwest. As an example, the Ohioan was perpetually disappointed in the intermittent monsoon rains, which never measured up to Ohio's spring showers. He announced the commencement of the rainy season three times that summer before finally grumbling on August 13, "I guess this is the commencement of the rainy season which is unusually late in beginning."[16]

Way's journal ends after four months. The homesick mining executive left Arizona after about two years. The fortunes of Way and Poston were related in that both of their designs ended with the outbreak of the Civil War and the removal of federal troops that had protected the mines from raiding Apaches.

Difficulties for the mining companies began with an Apache raid on a lumber camp near the old Canoa Ranch. This raid, followed by the betrayal of Cochise in 1861, sparked a new onslaught of Apache raids that plagued settlers in the valley for the next twenty-five years. The critical event that ignited the new wave of Apache violence was the decision by a boneheaded army lieutenant, George Bascom, to take family members of Cochise hostage in retaliation for a raid by White Mountain Apaches on a ranch in which a Mexican boy was taken hostage. Bascom took the "Apache" hostages even though Cochise assured him that Indians belonging to another tribe had the boy. Negotiations with Bascom were pointless, so Cochise began raiding and taking hostages of his own; the "bloody spiral," in the phrase of Thomas Sheridan, had begun.[17]

With the outbreak of the Civil War, U.S. troops in the valley soon left for the East. As the soldiers left, Poston recalled "The smoke of burning wheat-fields could be seen up and down the Santa Cruz valley." Mining became impossible without the protection of the army, and so once again Tubac residents abandoned the old presidio. Poston skedaddled, claiming to have left "over a million dollars" in company property behind.[18]

Both Poston and Way served in the Union army during the early years of the Civil War. But whereas Way never returned to Arizona, Poston soon wrangled a federal appointment as Indian agent. In 1864, Poston was back in Arizona, accompanying another of the long line of sojourner/journalists, J. Ross Browne. The duo traveled through the Santa Cruz River valley late in 1864. Browne's journal recorded observations of dubious reliability along with adventure stories of romantic proportions. Primarily a fanciful rendition, his account adds only a few tidbits to our understanding of conditions in the valley.

Regarding the jornada from the Gila to Tucson, Browne noted "the wells dug by the Overland Mail Company," and the seasonal pools near Picacho Peak. This offhand remark gives an indication of an early effect of commerce in the valley. The lower basin north of Tucson would be the last to see widespread development following World War II. This reference to wells at the Butterfield way station near Picacho is the first reference to the thousands of wells that one day would make agriculture possible in the lower basin.

Browne's observations at times ring true, but then a flight of imagination intrudes, and the intrepid reader is left to wonder what to accept. In one example, Browne described Tumacacori in plausible terms: "Tumacacori is admirably situated for agricultural purposes. The remains of acequias show that the surrounding valley-lands must have been at one time in a high state of cultivation." Browne was correct in his supposition, but his sketch of the mission showed no hint of acequias or pastures. Instead, the old mission was imaginatively rendered surrounded by saguaros. Clearly Browne was writing, and drawing, to suit the expectations of an eastern, drawing-room audience who already in the 1860s associated "Arizona" with cactus. Saguaro cactus occurs in a very restricted zone determined by climate and elevation, and has never flourished in recorded time in the Tumacacori area.[19]

The town of Santa Cruz did not impress the traveler. He noted, however, that the village was engaged in trade with the mines in the Santa Rita Mountains to the north. He attributed the trade with the Mowry Mine as sustaining the settlement. Browne noted that there was a small dry-goods store in the village, run by "a German Jew, named Apfel. . . . Flour, corn, and pinole may be had occasionally, but the supply is scanty and uncertain." Although Browne made no reference to it, Solomon Warner, who had fled to Santa Cruz during the Civil War, had married a Mexican widow in Santa Cruz and had established a mill, freight business, and small store in the town. Was "Apfel," the owner of the dry-goods store, actually Warner, the name changed by Browne in an effort to conform to an eastern reader's expectations? Since no stores had graced the village in the recent past, it seems highly unlikely that two stores would be operating in Santa Cruz that winter of 1864.[20]

The river's history in the 1870s clearly diverges between the upper basin, from the headwaters of the river in the Canelo Hills to the Canoa Ranch north of Tubac, and the middle basin, in the vicinity of San Xavier and Tucson. Unacknowledged by Tucsonans, the tide had turned in the Apache wars. No longer was the town itself threatened with Indian attack, although isolated miners and ranchers remained vulnerable to Apache assaults. Thus the middle basin grew in population and commercial activity through the 1870s, while the upper basin lagged in development until the Apache raids finally came to an end in the 1880s.

Mining in the mountains surrounding the valley in the upper basin continued through the 1870s, though limited by Apache raids and hampered by the difficult and rudimentary transportation system in place: concentrated ore had to be loaded onto wagons and hauled to ports in Mexico or to railroad links in the United States. Sylvester Mowry established one of the most successful mining operations in the Patagonia Mountains during this time, struggling against these difficulties for years. In an emblematic circumstance, given the expense of transporting the valley's ore, Mowry died in England in 1871 while on a fund-raising trip. The real boom in mining came in the 1880s, when railroad access spread across the territory, the Apache threat receded, and mining technology improved with the development of steam and electric power, more effective explosives (dynamite over black powder), and high-speed pumps to eliminate groundwater from mine shafts.[21]

Miners may have been the first to bring a truly modern perception of groundwater hydrology to the valley. But given the limited extent of mining in the 1860s, and the mines' limited demand on water resources, underground or otherwise, the heightened awareness and modern perception among miners had little affect on the river. A mine turning "wet," with shafts and drifts intersecting underground water supplies, would have been considered by miners either a nuisance or a catastrophe, depending on the amount of water in the shafts and whether pumps were available to remove it.

Mines affected the river primarily through timber cutting in the watershed. In a fashion similar to cattle grazing, while mining activity was limited, the ecological effects were limited. But as mining operations proliferated and increased in magnitude in the 1880s, the demand for timber by the mines increased to the point that ecological effects became highly probable, even if difficult to quantify. Hadley and Sheridan explained that "fuel-wood cutting" comprised the greatest ecological effect of mining in the region. Hillsides near mines and mining settlements soon became denuded of trees as workers and their families cut timber for both industrial and domestic use. As one observer noted in 1882, "The hills adjacent to the town [Harshaw], have been denuded of the . . . beautiful trees by which they were adorned, and the birds that were wont to sing to us . . . have departed."[22]

The river continued to flow in the upper basin in its old pattern despite mining and the return of ranching to the upper valley. As with mining, initially, livestock herds were too small to affect the river. Attacks by Cochise and his Chiricahua warriors in the area caused the abandonment of the San Rafael Ranch in 1865, thus the old pattern of development and retrenchment, ebb and flow, continued in the upper basin. It took about ten years for ranching to tentatively return to the valley. Six Mexicans from Santa Cruz, perhaps "*parcioneros,* or heirs of *parcioneros*" of the old San Rafael de la Zanja grant, were running cattle in the valley in the late 1870s. Two Americans, including Dr. Alfred A. Green, were also running cattle there. Green claimed to hold legal title to the entire grant. Rollin T. Richardson bought out these stockmen in 1880 in order to start the process of consolidating grasslands in the valley into a large-scale ranching effort. Richardson sold his ranching interests to Collin Cameron in 1883, and cattle herds expanded into the tens of thousands by the end of the decade. According to the 1870 census, the entire state of Arizona had only 5,132 cattle, although scholars suggest that the federal number is too small (James Wilson estimated the number to be 37,694 cattle). To illustrate cattle ranching's humble beginnings in the region, Henry Hooker established the Sierra Bonita Ranch in the Sulphur Springs Valley in southeastern Arizona in 1872. Hooker withstood Apache raids, and by 1880 had increased his herd to 5,500 head. Other ranchers, such as Pete Kitchen in the area of Nogales, accomplished similar feats, but all of their efforts remained constrained until the last Apache raids ended in the next decade.[23]

The river in the upper basin flowed without serious constraint into the 1880s. In the middle basin, the stirs of commerce and development brought water shortages into high relief. General indications of development included the establishment of Ft. Lowell in 1866, the arrival of the telegraph in 1873, the proliferation of mills and irrigation ditches through the 1870s, and finally, the arrival of the railroad in 1880.

The U.S. Army established Camp Lowell just to the east of Tucson in 1866 (about where the Santa Rita Hotel now stands).[24] Eventually housing about 100 soldiers, the post added to the population of Tucson, but not in such a way as to severely affect the river or general water supply. Rather the establishment of the post serves as another step in the development of a modern perception of the river. This is not due to army engineers or surveyors describing the aquifer in detail, but rather because army doctors and signal corps officers began recording daily temperature and precipitation readings. The army post created a data stream, from which the modern vision of the river emanated. For example, it is from this source that assertions about "normal" rainfall in the basin originate.

One of the first observations from the army post, made by John Spring in 1866, noted that the river "had a few places where the water was perhaps a little over four feet deep." The first government readings were taken during a wet cycle in the valley. Sam Hughes described the years 1868, 1869, and 1870 as among the wettest ever in Tucson. He recollected that during those years, the river flowed almost all the way from its source in the Canelo Hills to the Gila. Although that was no doubt an exaggeration— the river had not flowed over its entire course, except during spectacular floods, since the onset of semiarid conditions about 8,000 years ago—other accounts corroborate the wet years at the end of the 1860s. As a result of the increasing flow in the river, cultivation in the middle basin extended to Nine-Mile Waterhole, near the confluence with Rillito Creek. By 1869, more than eighty farmers cultivated almost 500 acres near the stage stop at that site.[25]

By 1870 the wet conditions had become "normal," at least to assistant surgeon Charles Smart, an army doctor visiting Fort Lowell. The doctor's report on the river valley was part of a general survey of army barracks and hospitals and the incidences of disease and illness. In Tucson, the concern about malaria along the river caused the doctor to make observations that validated the agricultural practices in the valley going back as far as the Hohokam. The doctor represents the circumstance of the newcomer in Tucson who takes the environmental conditions to be static and absolute. Weather cycles are generally too long in duration to be easily discerned by a transient population. On the other hand, although memories of long-term residents are valuable to get at the shifting climatological patterns, the vagaries of memory make them problematic. Fortunately, the growing availability of statistics and data provide additional sources to coincide with the oral traditions.

Doctor Smart described the river in Tucson as irrigating fields "for a distance of about three miles north and south, and on both banks of the river to the west of the town." The doctor confirmed the practice of growing two crops annually: "one of small grain, such as barley or wheat, sown in November and harvested in May, the other of corn, planted in June and harvested in October." The winter crops typically received only two irrigations, then relied on winter rain to mature the crops. The "malarial poison," resulted from the practice of flood irrigation in the summer, which inundated the fields near the town: "As cultivation can only be carried on successfully by irrigation, it follows that more or less of the fields are constantly under water, which, combined with the heavy rains in July and August, the tropical vegetation and its rapid decay, favors the development of the malarial poison, and accounts for the cases of remittent and intermittent fevers and diseases of the liver which prevail among the Mexican inhabitants during the months of August, September, and October." The doctor went on to remark that Fort Lowell, "being separated from these fields by the town, and being on a somewhat higher level, is almost exempt from these malarial visitations."[26]

Adding to the development of a modern perception of the river was a railroad survey in 1867. William A. Bell conducted the survey, describing the Santa Cruz as "a perennial stream" for its first 150 miles, until disappearing beneath the surface near "a spot called Canoa. It then flows underground almost to St. Xavier [*sic*] (twenty miles), and again reappears at a spot called Punta de Agua. Beyond St. Xavier it usually again sinks, rising for a third time as a fine body of water near Tucson, enriching a broad piece of valley for about ten miles around that town, turning the wheel of a fair-sized flour mill, and then sinking for ever in the desert to the north-west." The developing modern perception appeared in Bell's reference to the aquifer in his description of the river's disappearance north of Tucson: "I believe it flows over the bed-rock and under the drift which covers it for the remaining one hundred miles from Tucson to the Maricopa Wells, where a large spring—the waters of the Rio Santa Cruz, as is believed— comes to the surface and flows into the Gila." While not satisfying as a clear definition of a line sink/source, Bell's reference gives further support to the notion of growing empirical certainty about the river and its underground source. We might have expected this effort at geological accuracy from Bell, a fellow of the Royal Geographical and Geological Societies in England.[27]

With increasing commerce and development in the late 1860s, agricultural production increased in the middle basin. In 1854, Lt. John Parke had noted 300 acres in crops in Tucson. By 1870, the river allowed cultivation of almost 100 acres at San Xavier, with another 1,900 acres in production at Tucson. The increasing demand at Tucson had caused the river to disappear into a sea of pasture and fields.[28]

Another survey and map in 1871 reiterates Fergusson's description of a river disappearing into a maze of irrigation canals, but the 1871 map was drawn after a prolonged wet cycle had expanded the area of cultivation in the basin. John Wasson, Arizona's surveyor general, hired S. W. Foreman to conduct a survey of the river valley in the Tucson Basin. The population of the old presidio had grown to 3,224 in 1870, and no survey of the area had been completed since Fergusson's map in 1862. Foreman complained of the "Apache menace," which added difficulty to the survey. Miners and cowboys, and isolated surveyors, felt the Apache threat most acutely.

Foreman's map gives a good indication of the increase in cultivated acreage from Fergusson's map in 1862. Fergusson had noted on his map that fields north of the town were only planted when there was sufficient water—in other words, during periods of above-normal rainfall. Foreman's 1871 map clearly shows the northern fields and gives tabulations for the platted acreage in each surveyed township, including those north of St. Mary's Road and Congress Street. Whereas Fergusson had attached a label of marginality to the northern fields in 1862, no such qualification intrudes into Foreman's 1871 maps or notes.[29]

At about the time Foreman was surveying and drawing his maps, Solomon Warner returned to Tucson from Santa Cruz, where he had fled during the Civil War. Back in Tucson, Warner bought land near Sentinel

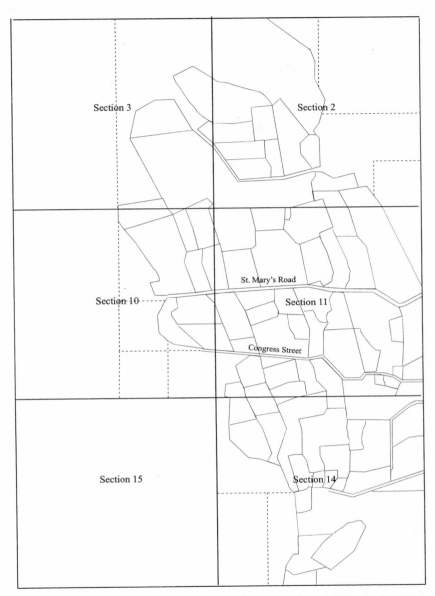

Section 3

Section 2

St. Mary's Road

Section 10

Section 11

Congress Street

Section 15

Section 14

S. W. Foreman's survey in 1871 indicated increased acreage under cultivation in the Tucson Basin from that mapped by Fergusson in 1862. Fergusson's map includes only the southern half of section 11, while Foreman's map includes all of section 11 and shows many more fields to the north in section 2. No label of marginality for the northern fields appears in Foreman's map. (Based on a map at the U.S. Department of the Interior, Bureau of Land Management, Tucson Office)

Peak and built another mill, acquiring water from an irrigation ditch that traversed a section of the old garden at the San Agustín mission. His mill competed with the two other mills in town, which had passed through several hands up to the 1870s. However, Warner soon became embroiled in legal suits over the construction of his millrace, and so accounts of the construction made their way into the legal records. Warner claimed the millraces were poorly constructed, and so refused to pay the builder. The builder then sued Warner for payment. The millrace for Warner's mill began near Rowlett's old mill, now called the Pioneer Mill, and ran over a mile to Warner's mill. The tailrace delivered water directly to the Acequia Madre east of the abandoned mission.[30]

More development schemes proliferated through the 1870s in Tucson. In 1879 Leopold Carillo began construction of a new ditch near the old El Ojito well. The ditch irrigated about twelve acres of formerly uncultivated land south of the road to the San Agustín mission. Carillo planted fruit trees on the land.[31] At Silver Lake, Richey and Bailey (to be followed by Maish and Driscoll) operated a resort of sorts, with a pavilion, hotel, and swimming and boating facilities. A tram service traversed the several miles from Tucson to the lake. Carillo established a water park of his own on South Main Street, called Carillo Gardens, in 1870. Warner constructed another small pond in the 1870s, catching water that flowed out of Silver Lake during the night. Warner's second lake furthered his legal problems in the 1880s as downstream farmers complained that the pond impounded public water. Development schemes even went beyond the river's floodplain. Carillo, then Tucson's most prominent landlord and developer, also announced plans to build a reservoir on his ranch at Sabino Canyon.[32]

In the meantime, domestic water supplies in Tucson were falling short. The old practice of relying on El Ojito, the artesian well to the west of the town, augmented by sale of water by wagon from a hand-dug well on South Main Street, had proven inadequate. The first plan to establish a water system was made in the spring of 1879. T. J. Jeffords, friend of Cochise and future recipient of movie and television fame in "Broken Arrow," proposed to develop a system to deliver artesian water to the city. Jeffords's plan never succeeded, but within three years construction on another municipal water system had begun.[33]

Water shortages had also developed at Fort Lowell. In 1873 the fort moved from the outskirts of Tucson to a location near Rillito Creek about six miles northeast of town. Army personnel had picked the site in part due to the abundant water supply, and the initial systems of hauling water by

Silver Lake resulted from the damming of several springs south of "A" Mountain in the late 1850s by William and Alfred Rowlett. The pond initially provided a constant water source to power mills, and by the 1880s, when this photograph was taken, Silver Lake had become a popular recreation spot operated by Richey and Bailey, followed later by Maish and Driscoll.

mule, constructing acequias into the post, and building windmills with storage tanks proved adequate into the 1880s.[34]

The proliferation of mills and irrigation ditches in the middle basin created increasing demands on the river. During the wet cycle leading up to the 1870s, however, the river experienced no noticeable decline. But as always, cycles turn, and the wet years turned to irregular rain and periods of drought. The joint circumstance of floods and brush fires plagued the Tucson Basin in the mid-1870s. Wet years and seasons generated bursts of vegetation, which then dried into perfect kindling during dry spells. The fearsome brush fires experienced in 1877 followed "a prolonged winter-spring drought." The dry winter and spring was followed by a haphazard monsoon in the summer of 1877. The summer rains started with a flourish in July, but then disappeared in August.[35] Grass sprouted, dried, then burned as wildfires swept the basin. To complete the cycle, with the land-scape denuded of grass and vegetation, the town experienced unusual flooding in 1878. A huge storm swept into the valley in July, causing all the washes and tributaries in the eastern basin to flood through the town. On the Rillito, floods washed out cottonwoods and set the stage for arroyoiza-tion that would follow in the 1880s and 1890s. That storm on July 11, 1878,

The tram from Tucson to Silver Lake traveled down this graded road for decades. Still a tree-lined lane in 1927, this photograph looks west toward the Tucson Mountains after a winter rain.

was in large part responsible for causing July of that year to be the second wettest month (5.72 inches) in Tucson's history (the wettest occurred in 1921).[36]

Climatological factors seem the most significant in these occurrences: the presence or absence of monsoons, as well as the occurrence, or not, of El Niño effects. The impact of overgrazing in the basin was probably limited. Sheridan described Hispanic ranching in the eastern basin, but its limited scope, still vulnerable to Apache raids, mitigated against severe overgrazing. An absence of groundcover may have exacerbated the flooding in 1878, but if so, the denuding of vegetation in the eastern basin was the result of a natural force—fire—not a four-legged grazing machine. As mentioned previously, when Fort Lowell moved into the eastern basin in 1873, water was plentiful in the Rillito. Through the 1880s and into the 1890s, competition for water grew as ranching and farming expanded. The development of wells and pumping technology in the 1890s finally solved the disputes over water in the eastern basin, but also lowered the water table and contributed to the arroyoization of the channel.[37]

The 1880s saw the blossoming of development in the Tucson Basin which finally and irrevocably started the decline in the water table. The transition to American sovereignty had not rendered immediate change in the river's flow, but did eventually bring to the river valley developers and engineers who looked at the river with a modern perception. Eventually that vision would reach critical mass, and the valley's water resources would feel the effect of constant and steady increases in utilization and manipulation. The variable of human settlement was increasing in weight. New, thirstier cultures had arrived in the valley—American, Oriental, and industrial. As these cultures took root in the valley in the 1880s, the river began to disappear.

PART II
Modern

8

Steel Rails and Steam Pumps

1880–1900

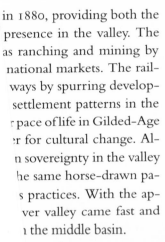

in 1880, providing both the
presence in the valley. The
as ranching and mining by
national markets. The rail-
ways by spurring develop-
settlement patterns in the
r pace of life in Gilded-Age
er for cultural change. Al-
n sovereignty in the valley
he same horse-drawn pa-
s practices. With the ap-
ver valley came fast and
n the middle basin.

lly unnatural, however?
process, but neither of
ly the river had dried up
o reappear and aggrade
ural practices had been
years, long before the first humans
ever set foot in the river valley. However, the accelerated pace of life in the
valley was new, and ultimately the railroad shifted the focus of the river's
history to a subsurface, aquifer narrative.

Along with the railroad came a new way of looking at the river—a
modern perception—viewing the water on the surface as only the tip of a
much larger underground supply. This perspective was not completely
new, since recent surveyors and engineers had looked at the river this way—
even the Hohokam may have viewed the river through a quasi-modern
prism—but the increasing prevalence of this vision marked the waning
decades of the nineteenth century. The spread of the modern perspective,

along with industrial technology's ability to bring such visions to life, heralded a distinct change in the way human inhabitants of the valley were to use the river.

The modern era is marked by a quickening pace of change. In the river valley, the middle basin underwent the most profound changes, but the other basins to the north and south also felt the effects of the railroad and industrial technology. Although the degree of change varied throughout the valley, industrialism's presence was so pervasive that no portion of the valley escaped its influence. The lower basin north of Tucson remained quiescent the longest—not even the railroad could make the dry, riverless area immediately attractive to farmers or ranchers. With the railroad came pumps, however, and eventually even the lower basin sprouted with irrigated agriculture. Likewise, the upper basin south of Tucson seemed initially to escape the effects of industrial technology, the river meandering in its ancient pattern through grasslands and floodplain agriculture. But the railroad in the upper basin lagged only two years behind its cousin's arrival in Tucson.

The modern era is also marked by an increasing faith in science and technology. It was an age for engineers, and this trend toward rationalism fostered a need for recording observations of and compiling information about the river. As the river began to disappear, scientists and engineers began creating data streams to track the process of decline, much to the benefit of later historians. To accommodate the proliferating source material, I will discuss the path of development of each basin of the valley separately; hopefully, this will serve to make clearer the river's changing circumstances.

The railroad extended its reach into the upper basin in 1882, two years after its arrival in the middle basin, with a round-about link between Tucson and Mexico through the canyon formed by Sonoita Creek. The presence of the railroad coincided with the declining Apache threat to facilitate the return of large-scale ranching to the upper basin. The railroad also stimulated the development of a renewed settlement at Calabasas, and the border communities of Nogales, Arizona, and Nogales, Sonora. The border towns in particular brought a shift in the settlement pattern in the valley, as a major population concentration occurred in a tributary canyon rather than along the river itself.

Ranching in the upper basin expanded slowly through the early 1880s as a few stock raisers sought to exploit the lush grasslands. A half-dozen

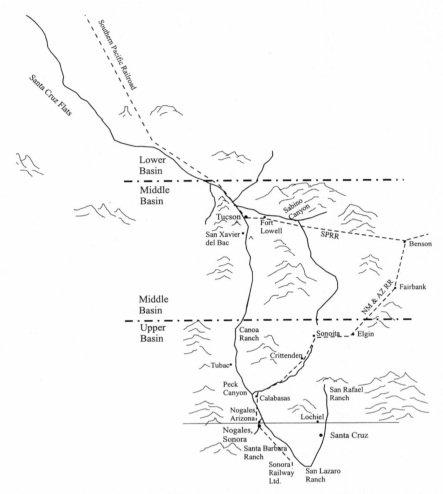

The arrival of the railroad and industrial society in the 1880s stimulated development in the river valley just as the threat of Apache raids began to wane. Nogales and Calabasas sprang to life as railroad towns, and Tucson received an economic boost as the railroad fostered increased ranching and mining in the region.

cowboys and small-scale ranchers, some probably descendants of the original parcioneros from Santa Cruz, were running cattle in the San Rafael Valley in the early 1880s.[1] Other ranchers filed claims in side canyons of the river south of Tubac. One such family included the Owen brothers, their sister Lucille, and her husband William Walker. The Owen brothers and the Walkers agreed that the grassy hills of southern Arizona resembled the hill country of California—good cattle country. The lure of the grass, and the little river, were so great that the family emigrated from northern

California, despite the understanding that Apaches still mounted occasional raids. Cowboys working isolated ranches remained most vulnerable. One of those vulnerable cowboys was Charles Owen, my grandfather's uncle, killed by Apaches in 1886 while working in Peck Canyon.

The resurgence of ranching took place on both sides of the border. The old Santa Barbara Ranch along the river south of the international boundary began its slow renaissance as the Mascarenas family consolidated thousands of acres. By 1890, the family controlled over 90,000 acres of prime grazing land and fertile bottomland along the river.[2]

Ranching took on a distinctly capitalistic flair in 1883 when Colin Cameron, and his brother Brewster, started the process of consolidating the old San Rafael de la Zanja land grant. Cameron was the head of an investment group from Pennsylvania, and came to the Southwest solely due to his perception of a great profit potential. As Hadley and Sheridan stated, "Unlike many of the Mexicans, Texans, and other Southerners [as well as Californians] moving onto the Arizona range, Cameron viewed ranching as an investment, not a way of life. He admired the tenacity of these pioneer stockraisers but believed that the small rancher was doomed despite 'his courage and his gun.' "[3] Clearly the Camerons brought a modern perception of the cattle industry to the valley, and so represent the new perception of the environment in the region, regardless of their understanding or concern for the river's aquifer. They sought to wring every ounce of profit out of the valley's natural resources of water and grass.

Colin Cameron aspired to control the entire San Rafael Valley, and exerted financial and political pressure to accomplish that goal. The Camerons' specific goal was to gain rights to the "overplus" lands beyond the grant's original 17,000 acres. Brewster was the lawyer who prosecuted these claims both in the courts and in territorial politics, and Colin supervised the process of denying homesteaders, or "squatters," access to the open range. One method Colin used was to place aggressive longhorn cattle on disputed range, effectively keeping the homesteaders' more domesticated cattle off the land. The expansion of the Camerons' herd was remarkable, and the stockmen pursued domination of the valley relentlessly. The herd had grown to 7,000 head by the spring of 1885, and increased to 17,000 by late 1887. Even though in 1900 the courts finally confirmed the size of the ranch to be the original four square leagues, or 17,361 acres, denying the assimilation of the overplus acreage, the Camerons' cattle had roamed over much of the overplus lands in the meantime.[4]

Colin Cameron's influence on the ecology of the valley went beyond increasing numbers of grazing cattle. He also supervised the development of water supplies in the valley, digging several wells and erecting a windmill. He also developed springs throughout the valley, piping the water to nearby locations.[5] In another example of the cattle company's ecological control, the Camerons protested a plan by miners in the hills to cut timber to supply Fort Huachuca, at the northern foot of the Huachuca Mountains, fearing that the loss of timber for this commercial purpose, as well as the added timber cutting needed to fuel the mines' smelters, would increase erosion by depleting the watershed of valuable plant cover. After complaining to the Secretary of the Interior, the woodcutting on the land grant ceased.[6]

The Camerons understood the workings of power in Washington and Phoenix. They also understood that the clearest manifestation of power in the river valley was through the control of water. In the competition with "squatters," the Camerons used the control of water as a tool to control the range. When independent stock raisers planned to move cattle into Harshaw and Mowry Canyons in the Patagonia Mountains, the San Rafael Cattle Company thwarted the move by developing water resources and wells in the canyons for their own stock. This theme of water and power runs through much of western history, perhaps most notoriously in California's central valley, where Henry Miller used every trick in the book, and invented some new ones, to gain control over more than one million acres. The Camerons would have felt a great affinity with Miller and all the other power elites who used the control of water as a source of power.[7]

Small-scale farming along the river continued during the Camerons' reign, probably in a fashion more-or-less constant since the Mexican period. Stock raisers routinely grew forage crops along the river, and a family nominally employed by the ranch planted about fifty acres near the Canelo Hills in "corn, beans and melon."[8]

The Camerons clearly understood the benefits of the railroad. Colin shipped his cattle to markets as far away as the East Coast. And as drought conditions set in at the end of the decade, Cameron also used the railroad to sell off almost half of his herd while the animals remained healthy and plump from previous seasons of bountiful grass. Cameron included some of his prized Hereford steers so as to bring in the best prices possible in the depressed market. In all, the ranch made five railroad shipments to California in the fall of 1889.[9] In that action, Cameron established a trend for

Arizona ranchers, raising cattle as young feeders, rather than tending them for three or four years. As ranch historian Jane Brewster explained, "Thereafter, Arizona ranches became essentially breeding establishments."[10]

The cattle boom of the 1880s lasted until drought conditions reduced the grass coverage on the range late in the decade. The trend was noticeable in 1885, but good grass years in 1887 and 1890 gave some ranchers a false sense of optimism. The drought hit with a vengeance in the 1890s, and the cattle industry in the Southwest crashed. Cameron was prescient enough to reduce his herds, as did several other of the large operators. Nonetheless, the range deteriorated, and many local observers, including Cameron, blamed overgrazing for the decline in grass coverage. The politics of class played a role in this analysis of overgrazing. Large operators such as the Camerons criticized small competitors for their ignorance and greed, accusing them of overgrazing the range knowing full well the damage they were causing. On the other hand, what choice did the small rancher have? Their margin of profit was so small, the alternative to overgrazing may simply have been economic oblivion, which was the fate most of them faced anyway.

Hadley and Sheridan cited climatological research that pointed to a combination of factors in the decline of the range. The drought conditions coincided with periodically heavy El Niño rains, resulting in huge floods that gouged deep arroyos, primarily in the middle basin. The formation of arroyos coincided with overgrazing in the upper basin, but a direct correlation is lacking. Cooke and Reeves, for example, posited that entrenchment of the river "arose from several interrelated environmental changes." Climatic cycles, changes in vegetation, and human machinations in the river channel all contributed to the formation of arroyos. As Hadley and Sheridan summarized, "Overstocking was just one of the factors that degraded Arizona ranges and watersheds. Droughts interrupted by intense El Niño storms also contributed to the devastation."[11]

The return of large cattle herds to the upper basin, and then their decline, marked a continuation of the traditional ebb and flow of development in the valley. Did community development in the upper basin also follow the same cycle of ups and downs? Just as the railroad could bring new towns to life, the absence of a railroad, or its demise, could spell doom for any aspiring community. Clearly, the ebb and flow patterns applied to the towns and villages of the upper basin as well.

The railroad found its way into the upper basin due to competition between the Southern Pacific and Santa Fe railroads. Each railroad company sought to link as much hinterland area to their main trunk lines as possible; in southern Arizona the railroad provided links for passengers and commerce, as well as freight service for the area's mines and ranches. In the Santa Cruz River valley the competition between railroads caused a temporary shift in the valley's main transportation route. The old wagon road followed the river south from Tucson to the border, but the first railroad link from Tucson to the border took a circuitous route east to Benson, then south and west to the border. A joint venture by the New Mexico and Arizona Railroad and Sonora Railway, Ltd. met at the border in 1882, traversing the 88-mile length of track from Benson, through Fairbank, Crittenden (later transplanted to Patagonia), Calabasas at the confluence of Sonoita Creek and the Santa Cruz River, and up the Nogales Valley (so named for the native walnut groves mentioned in the original Mexican land grant) to the border. By late spring, 1882, the renewed settlement at Calabasas numbered about 150 residents. The settlement spread onto the bluffs overlooking the river and included sixteen saloons, two dance halls, an opium den, five stores and one hotel. Nicknamed "Hell's Hollow" and rowdy, to say the least, the town's reputation grew with each shooting.[12] Nogales, Arizona and Nogales, Sonora, straddled the border, and within one year about 800 residents occupied the two towns. No doubt as rough-and-tumble as any railroad town, the residents initially occupied the riparian area in the canyon, and provided for their domestic water needs by sinking shallow wells directly into the floodplain. As the border towns grew, water systems became necessary, and one of the early pioneers, Leopold Ephraim, developed a private water company. The utility commenced operation in 1896 based in Nogales, Arizona, but controlled wells on both sides of the border. The company soon installed the first pumps in three of its wells in Mexico, and provided water to residents in both communities.[13]

Along with these new communities, the old village of Tubac continued to show signs of life. The farming of John Smith had prospered into the 1870s, and further efforts to exploit the surface flow of the stream at Tubac and downstream at Canoa continued. As a sign of increasing stability given the decline of Apache threats, residents, including Smith, organized a public school district, and in 1884 the federal government recognized the territorial townsite of Tubac. In 1885 residents built the first schoolhouse.[14]

The railroad entering the valley along Sonoita Creek serviced the mines in the surrounding mountains and gave life to the small communities of Fairbank, Crittenden, and then Patagonia. But just as railroads could breathe life into a community's aspirations for prosperity, it could also, by its absence, bring on stasis and decline. Such was the case of Santa Cruz, Sonora, one of the oldest communities in the valley and off the beaten path of industrial development. A railroad survey in 1906 marked a potential route to Lochiel, within a few miles of Santa Cruz, but that spur was never built. Eventually a rail line linked Santa Cruz to Nogales, Sonora, but the small community never seemed to benefit greatly from the railroad connection. The significant flow of commerce was crossing the border at Nogales, not running parallel to the border at Santa Cruz.[15]

In the middle basin, industrial development constituted a tidal wave of change in the river valley, and Tucson was at the crest of the swell. Some of the old patterns of development still persisted; for instance, efforts to enhance agriculture, and attempts to harness the river's flow to turn water wheels. These efforts focused solely on the surface flow in the river channel and so marked a continuity with the premodern culture. But the arrival of the railroad in 1880 ushered in the new modern perspective, complete with engineering works and speculative schemes that could, and eventually would, utterly dominate the river channel. Soon there was little left of the original meandering stream. Efforts to exploit the water resources did not end at that point, however, they simply shifted in focus from the modest surface flow to the huge underground supply.

Three major developments defined the transition to industrial culture in the middle basin. First, the continuing effort to benefit and profit from agriculture caused ever more grandiose engineering schemes in the river channel. In 1891, wells and steam-powered pumps brought completely irrigated agriculture to the basin, and one particular effort to engineer the stream resulted in the arroyo that still defines the river through Tucson. Second, court cases in the mid-1880s established the legal doctrine of prior appropriation in the basin. On the surface, the court decision seemed to benefit to a greater degree Anglo landowners upstream from the Hispanic landowners. In actuality, the court cases confirmed a growing reality in the basin. Success and wealth in traditional endeavors such as agriculture required access to water, and as the supply dwindled, access relied more heavily on money. The court decision did not turn so much on ethnic identities as on economic realities. The old system, harking back to the days of Jesuit missionaries and Pima farmers, distributed water to those fields that needed it most. That system was finally and completely swept away by court cases in 1885. The third major development in the middle basin was Tucson's water utility. The water company got its start as a modest, gravity-flow system relying on a headcut near Valencia Road on the spring branch of the river. By the 1890s, the modest system took on more modern aspects as the company sank its first wells and Tucsonans began relying on water delivered through a system of mechanized water pressure.

Agriculture in the Tucson Basin enjoyed few interruptions as the decade of the 1880s opened. Moist conditions brought more-or-less constant river flows and the decade began with little sign of water shortages. In addition to water in the river servicing crops and flour mills, recreation became another client of the river. Silver Lake advertised boats for rowing and sailing and a hotel along the shore of the lake fancied itself "the finest summer resort in Arizona."[16] The effort to develop water resources continued apace and in 1883 Solomon Warner began developing perhaps the last uncontrolled water resource in the basin when he constructed another dam and lake at the base of Sentinel Peak. The dam was intended to catch the flow from the springs at the base of the peak that normally flowed into the west branch, or mainstem, of the river. Warner's description of the area clearly indicated that a cienega lay at the base of the peak, at least during cycles of bountiful rain:

> the pond is situated at the foot of a mountain [Sentinel Peak] and there are several springs at the base; and for a considerable distance the water oozed out so much that the cattle to avoid it made a trail through the misquite [sic] bushes and over the rocks at the base of the mountain. Tullies and water grasses grew on all the land the pond covers with the exception of three or four acres on the south and east side. Beside the stream of water that continuously ran in the lowest part of the land which the pond covers there were other depressions where the water remained all the time. The whole land covered by the pond is damp and moist soil with the exception of the few acres above mentioned.[17]

Betancourt described Warner's new lake (about where the 22nd Street bridge is now located) as encompassing twenty acres, projected to be fifty acres in size when filled.[18]

Warner's new lake affected downstream farmers in the basin by diminishing the flow into the public acequia, and so in 1884 the water overseer opened the floodgates and started draining the pond. Eventually Warner arrived at an understanding with the downstream farmers, since it was obvious the undammed springs had augmented the water supply in the irrigation canals. Unfortunately for the farmers downstream, landowners upstream along the main stem of the river (the city water mains were located on the spring branch upstream and about a mile to the east) were also diverting water for their own agricultural uses, and this diversion eventually cut off the flow to the acequias.

In 1884, downstream farmers north of "the sister's lane" (St. Mary's Road) brought a lawsuit accusing upstream landowners south of the road with depriving them of water. The land north of the road in need of water was over 1,400 acres generally planted in grain. One of the plaintiffs in the case complained that the upstream diversions left "hardly enough to reach down there to give a horse a drink of water."[19] The defendant landowners, Leopold Carillo, Sam Hughes, and W. C. Davis, had incited the lawsuit by cutting off the water's flow earlier that summer. When confronted by complaints, Carillo had promised that "surplus" water would be available for downstream farmers in the future, but only after the land south of the road had been fully irrigated. The promise reflected a specious water doctrine to the farmers north of the road. The farmers realized "surplus" water might never materialize, especially as the upstream landowners expanded their agricultural production, and besides, they needed the water now. The notion ran completely contrary to the traditional practice of distributing water to all fields upon their need. C. A. Dalton, one of the plaintiffs and a former water overseer, explained to the court the traditional practice in times of shortages of providing water to the oldest fields prior to new lands recently placed under cultivation, regardless of their geographic position along the river. A primary factor in the case was the perceived affluence of the defendants and their practice, at least in one case, of renting almost 150 acres of land to Chinese truck farmers. Dalton considered the Chinese "gardens" to be newly cultivated, and additionally accused the Chinese farmers of "stealing" water from the public acequia "every day." Wheat and barley, Dalton reminded the court, required watering only once a month.[20]

Did racism play a role in these complaints? Anti-Chinese sentiment flowed in waves throughout the West in the late-nineteenth century, and it is safe to assume that some element of Dalton's ire was the result of the new farmers' ethnicity. But given the overall tenor of the plaintiffs' case, one may assume that any new fields drawing water away from older plots downstream, be they cultivated by Asians, Mexicans, or Anglos, would have been criticized with similar zeal.

The concept of prior appropriation came to bear in the judge's June 1885 decision. Judge Gregg ruled that the land north of the river would get water only after the "fields to the south had been fully irrigated, including the Chinese gardens."[21] Tucson courts were not alone in their consideration of water law in the West. Such disputes appear in legal cases from the days of the California gold rush, and had been percolating through the intermountain West ever since. At issue were two basic legal doctrines pertaining to the distribution of water: riparian, which provided water more-or-less equally according to the needs of all users along the banks of a river; and prior appropriation, which established a hierarchy of rights among water users. Under prior appropriation, an upstream landowner with a clear claim to water could divert the entire flow to his or her purposes. There was also a third system in the West that harkened back to the Spanish period, a hodgepodge of royal degrees and priestly interventions that were neither clearly riparian nor prior appropriation. The common theme to all the legal disputes in the West beginning in the 1850s was the drift toward prior appropriation, and the judge in Tucson conformed to this pattern. Environmental historian Donald Pisani attributed these decisions and the move toward prior appropriation to "the nature of the local economy," which by the 1880s in Tucson clearly exemplified developing industrialism.[22]

Engineering works continued in the middle basin river channel as Tucsonans sought to maximize the amount of water available for agriculture. A common way to tap the shallow aquifer and obtain a surface flow for irrigation was to dig a shallow ditch that intercepted the flow of water just below the surface. The "artificial headcut" would capture the subsurface flow of the stream, and during high-flow periods, floods would enlarge and deepen the ditch, providing water during low-flow seasons. One such ditch provided water for the Canoa Ranch in 1887. The new owners of the ranch, Maish and Driscoll, originally planned to build a canal all the way to Tucson so as to irrigate land near the city. Although the grand scheme never came to pass, the first mile of the canal was built and continued to provide water for the ranch well into the twentieth century. The Santa Catalina Ditch and Irrigation Company hoped to achieve the same result by constructing systems on the Rillito in 1886. The company built canals in 1886, 1887, and 1888, but in each case the canals, originally two miles in length, were filled with sand during floods.[23] An artificial headcut was also the notion behind the ditch commissioned by Sam Hughes near St. Mary's Road in Tucson in 1888. Hughes initially sought to provide water sufficient to irrigate 15,000 acres with a ditch twenty feet wide in 1887. A shortage of funds delayed the ambitious project, and the next year Hughes constructed a more modest ditch, relying on the seasonal floods to enlarge the heading of his diminutive canal.[24]

Initially, Hughes's ditch made little impact on the river channel. "Minor flooding" in October 1889 started the erosion of the ditch, but it wasn't until the floods of July and August 1890, that the headcut began "taking a walk to Maish's lake [Silver Lake] for water." As the newspaper reported: "The 'raging' Santa Cruz continues to wash out a channel and the head of it is now opposite town. It may reach Silver Lake before the rainy season is over; who knows?" By the end of the month, the paper confirmed its prophesy: "The head of the new channel of the Santa Cruz river is now opposite Judge Osborne's place, on the road to Silver Lake."[25] The floods of August had damaged crops in the upper basin and washed out railroad bridges between Nogales and Tucson. Old-timers in the city proclaimed the river to be higher than at any time in the last twenty-five years.[26] Erosion from the Hughes ditch continued to expand, swallowing up fifty acres of land in the formation of the arroyo. The newspaper fretted over the arroyoization as damage to land in the basin increased: "there are now several channels being cut by the flood, all of which run into the main channel. If the flood keeps up a few days longer there will be hundreds of acres of land lost to agriculture. As these new channels or washes are spreading out over the valley, they will cut through and greatly damage the irrigating canals." The newspaper went on to call for efforts to prevent further damage.[27] The floods had continued throughout August, and on the twenty-sixth the newspaper again called for steps to be taken to prevent further damage from the river's "ravages . . . next summer."[28]

Perhaps inevitably, given the profound changes in the floodplain and the amount of property damage, the blame game soon began, even as the newspaper attempted to put a positive spin on the development: "The big arroyo cut through the Santa Cruz valley will afford the means of drainage for the city. It is an ill wind that does not bring good to some one."[29] Whose fault was the arroyo? Was the erosion of the channel caused by overgrazing in the upper basin and along the banks of the river, or by Sam Hughes tampering with the river channel? Opinions varied. Many observers blamed the Hughes ditch, but in 1899 Volney M. Spalding, a scientist at the Smithsonian facility in Tucson, cited the overgrazing argument: "Previous to the advent of cattlemen some 20 years ago, and the destructive effects of over-pasturing, the valley of the Santa Cruz had a luxuriant growth of saccaton and other vegetation, which prevented the cutting of channels,

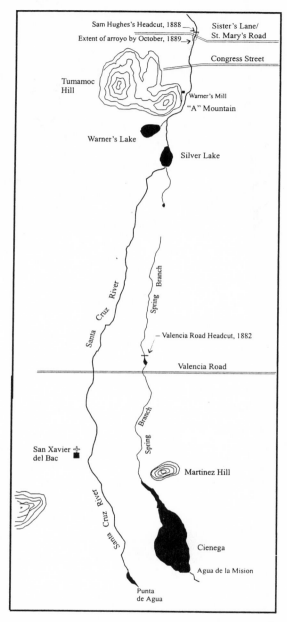

This map shows some of the historic landmarks in the Tucson Basin in the 1880s. The site of Silver Lake and Warner's Lake are clearly shown south of "A" Mountain, as are the two branches of the river. Sam Hughes's headcut had started the process of arroyoization near St. Mary's Road, but as yet the river remained shallow and meandering throughout most of the basin. (Based on John T. C. Parker, "Channel Change on the Santa Cruz River, Pima County, Arizona, 1936–1986," USGS Open-File Report 93–41, 1993, p. 10)

The flood of October 1889 began the erosion of the Sam Hughes headcut in the river channel. The formation of the arroyo through the Tucson Basin started here as the flood widened the cut, which started eating its way upstream. The water cascading into the ditch from the far side of the floodplain gives an indication of the way floods in the basin formerly spread out over the floodplain in a thin sheet a mile or more wide. After this, floods became encased within an ever-deepening and lengthening arroyo.

and the water spread out over the whole valley instead of flowing through the deep cuts it has since made; tules grew thickly in the springy places, and a fine forest of mesquite covered the ground." In a footnote, Spalding referred to the overgrazing argument as "the commonly received version," then went on to report the other view, "that about 20 years ago [the Hughes ditch was nine years old in 1899], certain old settlers undertook to 'develop water' at a point about 2 miles down the river." Spalding also noted that the entrenchment of the river would have the effect of lowering the water table.[30]

Within ten years the analysis of the entrenchment of the river had itself become entrenched. Olberg and Schanck in 1913 cited both causes, blaming overgrazing for changing the nature and the destructiveness of floods, and blaming the ditch digging in the river channel for starting the process of entrenchment: "when floods occurred the water spread out in a thin sheet over the valley, doing no damage . . . it did not run off as quickly, hence the floods, while they lasted longer, were not so great in volume and their principal effect was that of a thorough irrigation. As late as 1873 the valley was covered with grass. Shortly afterwards cattle were brought in, later the range was overstocked, and now it is nearly bare. The result is that when a flood spreads out over the valley to-day [sic] it seizes upon every cow path, wagon road, and other slight depression, and miniature torrents are formed which cut channels in every direction." On the other hand, the main arroyo had been caused by efforts "to develop ground water" by digging "open trenches which intercepted the ground-water plane." As Olberg and Schanck admonished, "Apparently no effort was made to protect the upper end of the trench from erosion, with the result that barrancas [ravines] were formed, which were rapidly enlarged by each succeeding flood." G.E.P. Smith blamed settlement practices, including overgrazing, for the entrenchment of the Rillito in the 1890s. In this he basically echoed Olberg and Schanck's analysis of the background causes of arroyoization: "The general effect of settlement was to increase the magnitude and severity of the floods."[31]

Betancourt and other recent scholars tend to support the Hughes theory. While it seems clear that the floods of 1890 were particularly large and destructive, it is doubtful the arroyo in Tucson would have formed without the presence of the artificial headcut made by Hughes.[32]

The arroyoization of the river caused reorganization of the irrigation system, which remained based on the old acequias serving the floodplain since Spanish days. The old canals were at the level of the shallow streambed, however, and now that the river flowed at the bottom of a deep channel, a new system would have to be developed. As it turned out, the arrival of industrial technology was already providing the solution to the problem, and the new system would be "Better than a Canal." The first farms in the basin dependent entirely on "the medium of pumps" started early in 1891 to great fanfare.[33] The recognition quickly dawned that pumping water from "inexhaustible" wells would "add wealth and population to this section of Arizona." At last the pinched limits to growth in the region dictated by the shallow flow of the river would be banished. As the newspaper proclaimed, the "duplex Portuguese pump" was a great improvement over the traditional irrigation system. The modernist perception appears in the newspaper's clear expression of frustration with the river's heretofore meager bounty: "Where water can be had at from 20 to 40 feet in depth in this valley, there is no reason why pumps . . . should not take the place of a small natural supply of water taken from the Santa Cruz river at this time." Pumps would provide much more water than "the insignificant supply of water furnished through illy [sic] constructed ditches."[34]

The evolution of this faith in pump technology was borne out in microcosm at Fort Lowell on Rillito Creek. By 1885 farmers and ranchers upstream from the fort began drawing water to such an extent that the old system of ditches and windmills began to fail. Wells were dug to 150 feet, but the windmill technology of the time made water levels below thirty or forty feet inaccessible, and so the post remained in competition for the limited surface flow in the creek. In the search for a solution to the water shortage, the army studied the possibility of constructing a dam and reservoir in Sabino Canyon, and in 1886 annexed the canyon to the military reservation. Nothing came of the plan, and the post's water problems were not solved until the acquisition of a steam pump in 1887. The pump provided access to the bountiful groundwater supply, and the fort's competition with neighboring farms and ranches for scarce surface water came to an end.[35] So the pump technology that allowed for the first fully irrigated farms along the Santa Cruz River in 1891 had served a few years earlier the fort and settlers in the eastern basin.

More plans to rehabilitate agriculture in the basin came about in the 1890s, and these efforts clearly included an effort to tap the aquifer. The Allison brothers, Frank and Warren, commenced an effort in 1892–1893 to capture the water source at the base of Sentinel Peak (Warner's new dam and lake had been washed away in the 1890 floods), and carry the water through an extensive canal system to fields to the north and west of the hospital road (St. Mary's Road). Significantly, the Allisons also sank wells at the head of their canal to a depth of ten to fifteen feet so as to augment the supply of water to their system.[36] No longer was the surface flow alone sufficient. By January 1893, the canal was feeding water into a reservoir covering about ten acres. The brothers planned to turn land currently used for grain production into fruit orchards, but quickly the Allisons' land had become too alkaline for cultivation.[37] The brothers constructed another canal on the east side of the river, eventually delivering water to fields north of Tucson. The wells continued to gush extravagantly and the brothers coined the name for their project and farms "flowing wells." The water also powered a flour mill. Warren Allison later recalled, "This land we cleared and planted in alfalfa. We farmed it for several years and finally sold it . . . for $60,000." The new owners extended the ditch south to San Xavier and continued to develop the land west of St. Mary's hospital, which eventually included a dairy.[38]

Dreamers, visionaries, and con men announced other plans. In the Tucson Basin plans for canals continued to dominate the thinking of most developers, but forevermore canals in the basin would have at their source wells and pumps, not simply a gently flowing surface stream. In 1892 the newspaper called for "Home Enterprise," based on the utilization of groundwater, which promised to triple the amount of arable land in the basin.[39] In a variation on this modernist theme, another plan, presaging later Bureau of Reclamation projects, called for damming the river near Nogales, which would provide water for the irrigation of 300,000 acres from the border to the San Xavier Indian Reservation. Most of these plans never came to pass, but they indicate the transition to the modern perception of a virtually limitless supply of water in the aquifer.[40]

The system for delivering water to residents of Tucson went through a significant transition in the last decades of the nineteenth century. Throughout Tucson's early history, water had been drawn from wells in and around the town, and as the community had grown through the nineteenth century, this system remained relatively unchanged. A public well served all residents, and Tucsonans dug private wells on their own property. One variation on this simple system was the small business operated by Saunders and Phy in the 1870s. Entrepreneurs developed plans to bring water into the city through a gravity-flow system in the late 1870s. Then Mayor Robert Leatherwood supported these plans, and eventually convinced two Midwestern investors to finance the plan. The Tucson Water Company resulted, a privately owned operation that piped water into the city from a spring six miles upstream at the Valencia Road headcut. The water traveled most of the way in a redwood flume in the river channel, and made the final traverse into the city through a pipe constructed of "sheet metal coated with tar." The water mains started delivering water to Tucsonans in 1882. With the flow of water into the city, prospects brightened, and boosters proclaimed "that Tucson would fast become to Arizona what Denver is to Colorado."[41]

The water company's draw of water upstream became a factor in the shortages that developed for downstream farmers later in the decade. Solomon Warner accused the company's system of drawing off so much water that Silver Lake was lowered to the point where insufficient water remained in the millrace to power his flour mill.[42] Over the years and decades to follow, the city's draw of water would reach such levels that agriculture would disappear from the basin. But for the time being, ebb and flow cycles persisted, and Warner's perceived shortfall faded from memory as floods and wet cycles saturated the water table.

But drought conditions were never far away. The dire predictions in 1890 of further damaging floods failed to materialize, and the heavy rains of the summer of 1891 were replaced by drought conditions in the fall. By October the newspaper was describing a new set of gloomy circumstances: "The Santa Cruz river has never been as low since 1872 as it is now. At that time the people had to dig in the bed of the river for water, and barely obtained enough for home consumption. It is feared that the same conditions will come to pass this fall."[43] Fear no doubt visited Tucson that autumn, but the situation in 1891 was no longer analogous to the 1870s. A system of water mains served city residents, and although water shortages plagued the system, the accouterments of industrial society spared residents the necessity of digging for water in the dry river channel.

The development of the water company also foreshadowed a shift in the water politics of the basin. The water company primarily serviced domestic and industrial users, and so the company's policies eventually came to forsake agriculture completely. As an early indication of the urban focus of the company, and in a simple effort to maintain good public relations, the water company made an agreement with the city council in 1888 to encourage the planting of trees in the city. Anyone who planted a tree along one of the city streets would receive free water for irrigation.[44] In another manner of public service, the city council installed a public drinking fountain on Church Place: "It will be of cast-iron and have conveniences for both man and beast."[45]

No fashion of public relations could overcome the drought, however, and as the dry conditions reduced water supplies in the 1890s, the water company began developing wells to augment the city's supply. City customers complained that the water mains ran dry at times and residents had to wait until the late morning or early afternoon to receive any water through the pipes. Although the water company was still owned privately, the city council displayed an awareness of the water shortages in 1892 with a debate on making conservation measures mandatory. The council did not take any action at that time, but the mayor did order a halt to watering the grass in city parks.[46] In response to the complaints and the water shortages, the company began developing its first wells in 1893.[47] The well field was actually on the San Xavier Reservation, near the head of the existing water main at the headcut near Valencia Road. The projected depth of the well was twenty feet, and "the new plant" would include a pump capable of delivering one million gallons of water a day. The system finally included five wells within a radius of fifty feet, and went into operation at the end of

March 1893 with the hopeful sendoff: "Residents in the eastern end of town will have no cause hereafter for grumbling."[48]

The water company intended the new well system, in part, to alleviate residential complaints. But agricultural uses were never far removed from plans to augment Tucson's water supply. In April 1893 the newspaper reported that the new well system would also provide sufficient water to irrigate 640 acres of land owned by the water company. The company also planned to sell surplus water to area ranchers. This intent ran counter to the view, reported in the newspaper, that "the supply [of water] in the valley will not last the summer through." The projection seemed to be coming true in June, when residents complained that no water was available in the city's mains before eight or nine in the morning. Even in the midst of water shortages, however, the newspaper continued to boost the city and its prospects. In the same issue as the complaints about low and nonexistent water pressure, the paper ran a description of the valley as a prime site for sanitariums. As the come-on proclaimed: "land is cheap and good water can be found a few feet below the surface of the ground. If near the river, it is unnecessary to dig because the waters of the Santa Cruz give an almost inexhaustible supply."[49]

Given the accelerated pace of human machinations in the river valley, were there still natural occurrences that rivaled human efforts in their impact on the river valley? And how did the modernists in the valley view these natural circumstances and phenomenon? The floods have already been mentioned, and monsoons lay at the heart of the most damaging floods. The summer monsoon in 1886 produced a flood that surged down the Santa Cruz River from the upper basin through Tucson, causing $4,000 damage to the dams around town. More floods occurred in the monsoon of 1887. In July the newspaper reported that water was so deep in the river that "a mammoth steam boat" could be floated. Another even bigger flood occurred in September, and the newspaper reported "for the first time in many years, it is navigable from Tubac to the Gulf."[50]

As if to accentuate nature's power and dominance over human contrivances in the region, an earthquake estimated at 7.2 on the Richter Scale rumbled through the valley on May 3, 1887. The earthquake apparently disrupted portions of the aquifer. Several springs and wells around Tucson dried up, while others experienced sudden discharges of water. The source of the spring branch of the river was displaced about one mile to the south, upstream from the original Agua de la Mision. Before the monsoon arrived, the Indians constructed a dam at the new source of water. Given the nature of the summer rains, the Indians' new dam washed out and had to be rebuilt after the monsoons ended.[51]

Clearly, natural forces continued to show their strength even as human contrivances in the river valley increased in sophistication and intrusiveness. No matter how intricate the engineering, human artifice remained bound within natural parameters in the semiarid bioregion. If anything, the fear of insufficient water supplies increased through the early years of the twentieth century. An answer for the concern arose in the form of progressive optimism in modern technology. A faith in the science of hydrology and the modern technical expertise of engineers would see Tucson and the Santa Cruz River valley through its dusty purgatory.

An early example of the-future-will-solve-our-problems mentality occurred in the last decade of the nineteenth century. The use of steam power to drive the increasing number of well pumps created one of the first energy crises in Tucson. The heightened use of wood to fuel the pumps was exhausting the supply, and projections of doom began to circulate. Gone were the days when trainloads of mesquite could be shipped to southern California.[52] When the newspaper interviewed a wood supplier, however, the calming voice of progressive reason arose. Mr. Shortridge, "in the wood business," explained to the newspaper that, yes, the supply of wood was growing scarce, and would run out "at the present rate in five years," but by then, Shortridge explained, the San Xavier Reservation would commence supplying Tucson to meet its needs, albeit at higher prices: "The San Xavier reservation has a fine wood supply. This the Papagos are becoming aware of, and are raising prices accordingly." The Indians' supply of wood would be exhausted in another five years, but by then, hopefully, coal from "the fields of New Mexico, Colorado, and San Marcial, Mexico" would be available on "the new railroads."[53] By 1910 the projection came true as wood supplies diminished and prices rose, but instead of coal it was gasoline engines, and then electric power, that fueled the well pumps.[54]

A similar type of salve for the water-conscious boosters of Tucson occurred after the turn of the century. The mantra became "additional sources," and the city government and water utility embarked on the path of development beyond the valley, and ultimately, beyond the watershed. Growth depended on water, and the city's health required ever-increasing quantities, while at the same time the river sank inexorably beneath the surface of the valley.

The perennial flow of the river ceased to provide sufficient water for all Tucsonans by at least 1884 when farmers north of St. Mary's Road ran out of water. A few months before the water trial began, the city's engineer, J. P. Culver, summarized the amount of water available from the various springs in the basin, explicitly referring to "live water," or "visible" water, distinct in his thinking from the unseen and largely untapped underground supply. Despite the archaic and charming reference to water as living, Culver's report clearly indicated the increasing prevalence of a modern perception of the river:

> The water supply afforded by the Santa Cruz river I have for several years observed and can state with accuracy the amount visible at the dryest [sic] seasons of the year for the past three years, which is a crucial test to its value. The visible waters of the valley, of greatest quantity and value, in this locality, commences about nine miles southerly of Tucson, at the Punta del Agua near San Xavier, on the Papago reservation, where a gauging shows in the neighborhood of 700 miner's inches[55] of live water, which, with moderate development, could likely be largely increased. Following down the valley this water all disappears by sepage [sic], only a moderate, or a partial use being made of it for the cultivation of small tracts of land and some minor domestic uses. Six miles south of town the Tucson water company have [sic] developed about 170 acres of miner's inches of sweet and pure water. . . . About a mile below this point live water again makes it[s] appearance and gauging made in 1881–2–3 resulted in showing about 23 miner's inches. The next point below and following the line of water all the way is Lee's mill [Silver Lake] where gauging made in 1881 and 1882–3 [sic] showed from 500 to 700 miner's inches passing through the waste flume. An examination at Warner's mill race, the next point below, was found to carry about the same volume of water as at Lee's mill.
>
> The final point, and the last place examined in the valley below Tucson, where live water is diverted from the valley, is the lower settlement about four miles from Tucson, where the irrigation ditch carries about ten miner's inches. In the aggregate it is safe to assume of the total 1,403 miner's inches of water that there can be placed implicit reliance upon at least 1,000 miner's inches of visible flowing water during the dryest [sic] of seasons. . . . The live water as enumerated, so far as a careful study of the physical features of the valley are

concerned, looks to the conclusion that it is only a very small part, compared to the underlying water measures passing downstream through the coarse gravel, invisible to the eye, but by [d]igging it is soon reached, rarely being but a few feet below the surface. When reached, it is found in great quantity and over a large area in width. . . . I assume, without a shadow of doubt, aided by these observations, that there is an immense invisible subterranean water course in the water channel of this river; that in the aggregate would be astonishing, properly collected and developed.[56]

At his most conservative estimate of 1,000 miner's inches of water "during the dryest of seasons," Culver's report indicated that the river provided about one acre-foot of water every forty-five minutes, or about thirty-two acre-feet per day. This is the first effort at quantifying the volume of water in the basin, and as such serves as the modern bookend to the army lieutenant's temperature reading in 1854. Of course, it also shows just how modest the surface flow of the river could be.

As Betancourt described, "By the 1880s demand for irrigation and domestic water exceeded the supply afforded by the perennial flow of the Santa Cruz." Agricultural production would increase through the decade, but as more acres came into production, the "visible" and "live" water supply vanished. Ever-longer canals and ever-deeper wells kept the farming economy going in the basin, but the archaic pattern of water resource use in the valley came to an end. Indians still irrigated 1,580 acres of land on the reservation near Tucson and produced thousands of bushels of wheat and tons of barley hay, but became increasingly engaged in stock raising and copper mining in the early years of the new century. The future of agriculture in the valley was to be found beneath the ground and required large capital expenditures to exploit.[57]

Later hydrologists would posit that 1940 marked the loss of balance in the basin between the pumping of groundwater for domestic and irrigation purposes and the natural recharge of the aquifer. But since the water table began dropping in the 1880s, the loss of balance in fact occurred much earlier. The rate of overdraft remained small compared to later levels, and the water table remained near the surface for many years to come. But it was sinking, and the future of agriculture in the valley was inextricably tied to groundwater mining. Wet cycles and monsoon floods would occasionally recharge the permeable alluvium under the floodplain and bring the river back to a modest flow, now at the bottom of a lengthening and deepening arroyo. Wet cycles would also bring springs temporarily back to life, such as the seep at the base of "A" Mountain. From the moment of complete recharge, however, the churning pumps would begin the overdraft anew, and the process of depletion would recommence. In a telling observation made in 1912, government scientists Olberg and Schanck referred to the Santa Cruz River as "the so-called Santa Cruz River."[58]

From its head in the Canelo Hills far to the south, the "so-called" Santa Cruz River had undergone significant changes in the last two decades of the nineteenth century. Overgrazing and drought conditions had denuded the landscape in the San Rafael Valley, contributing to erosion in the high valley and downstream as well. The railroad had fostered the growth of new communities and the rebirth of old settlements. The stream emanating from Nogales Wash began to diminish as domestic use in the twin Nogaleses increased. Further downstream, agricultural and domestic use of water at Calabasas and Tubac increased. All of these placed increasing demands on the river.

The most visible signs of change occurred in the middle basin. The deepening arroyo and proliferation of wells had lowered the water table and the river's disappearing act commenced anew.

The lower basin, however, remained somnolent. A few wells served stage stops and later railroad water towers, but the river had not flowed through the low desert in aeons and the proliferation of irrigated agriculture remained decades off. As the nineteenth century closed, sand, rocks, and creosote dominated the flats, with only the occasional flood bringing the river channel to life.

The twentieth century witnessed a continuing round of schemes and plans to squeeze more water out of the disappearing river. Archaic views of the river remained, but the modern perspective now ruled. Farmers and urban engineers looked at the river and its aquifer as a natural resource to be used to its fullest extent. Water unused was squandered, whether in rivers flowing wastefully to the sea, or in underground reserves resting dormantly beneath the ground, and so the underground aquifer became the target of ever-more-sophisticated extraction techniques.

9

Water Mining and the Wagging Urban Tail

1900–1930

PROSPERITY IN THE VALLEY WAS STILL tied to agriculture at the turn of the century; farming, ranching, and to a limited degree, mining, determined the use of water. By far the largest draw on the water supply went to farms, and the arrival of industrial technology only furthered that circumstance. Tapping the aquifer allowed farming to move away from the river's flood-plain—particularly in the lower basin—and therefore greatly increased the amount of acreage under cultivation. In other desert regions of the West Bureau of Reclamation projects brought dry land into production. In the Santa Cruz River valley no such federal project came to the rescue of farmers in the early decades of the twentieth century. None was needed initially, although of course the federal subsidy of cheap water would have been gladly accepted. But unlike the aquifer beneath the central valley of California, which began to run dry in the 1930s, the Santa Cruz River's aquifer beneath the lower basin was just beginning to be tapped signifi-cantly in the 1930s. Decades of steady pumping lie in the future. The surface flow of the river had never allowed extensive agriculture beyond the Tucson Basin, since the stream rarely extended into the flats. The stretch of river through the lower basin had remained a mostly dry jornada for centuries. This was all to change as irrigated agriculture became possi-ble in the twentieth century.[1]

In the upper basin, ranching followed the return of grasses as drought conditions lessened after 1903. But the shifts in emphasis from patterns es-tablished by Colin Cameron remained slight and the overall conditions of the river's watershed and headwaters in the San Rafael Valley indicated a continuity rather than marked change. Cattle grazing on the hillsides and modest farming along the floodplain continued in patterns long established.

Urban areas were increasing as well in the early decades of the twentieth century, but the towns and cities in the valley used relatively little water compared to agriculture. Even though slight, the cities' draw on water supplies became the focal point for loud, contentious, and problematic water politics in the valley. As the water table began to drop, especially in Tucson, the reality of intractable water politics came home to roost.

Two basic realities confronted the growing community of Tucson as the new century dawned on the middle basin. The first concerned the source of Tucson's water, or sources, and the quantity that seemed constantly destined to fail to meet Tucsonans' needs. The other harsh reality facing middle basin residents was the carrying capacity of the water utility's delivery system, and the utility's inability at times to meet the peak demand of customers during the summertime regardless of the overall availability of water. These facts of life in Tucson are now shared by residents of virtually every other southwestern city, although the particulars of water politics vary from place to place. Mike Davis, Norris Hundley, and Marc Reisner, to name a few, have remarked on the centrality of water, and the politics surrounding its management, in the semiarid Southwest. Often contentious and divisive, Tucson's water politics are illustrative, in turn, of extremely naive optimism and remarkably unabashed cynicism.[2]

In the upper basin at the turn of the century, the headwaters of the Santa Cruz River continued to travel through a terrain that beckoned to cattle ranchers with the promise of bountiful grass, even if the current drought conditions made the promise somewhat ephemeral and illusory. Ranching in the valley continued to be dominated by the enterprise centered on the San Rafael land grant. In 1903 Colin Cameron sold his ranch for $1.5 million to William C. Greene, who had made a fortune in mining and ranching in Sonora. Greene consolidated the San Rafael Ranch with his holdings in Mexico, and eventually established one of the largest herds of registered Hereford cattle on the ranch. Greene used the Arizona property to raise breeding stock, while running feed stock on the ranches in Mexico, a practice which definitely lightened the demand on grass in the upland valley. Whereas the Hereford herd in the San Rafael Valley was about 5,000 head, with 500 registered Herefords, over 30,000 calves were born on the Mexican ranches in 1904.[3]

As to the impact on the river, Greene's ranch continued long-established practices, including the cultivation of farmland along the Santa Cruz River. Farmers grew vegetables to sell to the ranching community, and Greene used water from the river to irrigate forage crops for the livestock of alfalfa, corn, and milo maize. A major innovation during Greene's control of the ranch was the institution of widespread fencing, as farm plots and pasturage were fenced off. As Hadley and Sheridan noted: "the ranges were crossfenced into a series of smaller pastures, which allowed particular areas to have rest periods without grazing." The ranch managers also continued Cameron's practice of opposing woodcutting on the ranch and surrounding hillsides, seeking to preserve the watershed and avoid erosion of topsoil. Greene also experimented with dry-climate grasses, which he imported from Russia and Turkestan.[4]

As the ranch became a more exclusive breeding operation, the managers formalized the system of fenced pastures. The plots were designed to hold up to fifty head, and were arranged as to allow access to water in the river channel. If the small ranges had no river access, ranch workers piped water to them. Eventually the herd increased to 1,400 registered cows, which Hadley and Sheridan described as "possibly the largest registered herd in the country."[5] Development and use of the pastures were refined over the years to include permanent irrigation and rotation between grazing and the cultivation of forage. Over the ensuing decades, the efficiency of ranching in the valley became well known, especially on the Heady and Ashburn Ranch in the San Rafael Valley, which drew the attention of ranching officials and journalists (to be discussed in chapter 10). Of course, one reason for the continued success of ranching in the upper valley was the proximity to the headwaters of the Santa Cruz River. In the ancient pattern, the river provided the means for human subsistence.[6]

Another difference between Greene's large cattle operation and Cameron's was the policy regarding the smaller ranches on the borders of the land grant. Whereas Cameron had opposed these smaller operations— often ruthlessly—Greene dropped this opposition and never made an effort to claim the so-called overplus lands. As a result, homesteads and dryland-farming operations proliferated in the early 1900s, with farmers clearing land of trees and woody shrubs primarily to cultivate beans and corn. This spread of agriculture in the valley took place within a context of at least partial government regulation, however. The creation of forest reserves and grazing allotments mitigated somewhat against rampant overgrazing and over-development in the valley. As Hadley and Sheridan summarized: "The San Rafael Valley has undergone less ecological damage than other similar river valleys in southern Arizona."[7]

The valley may have fared better than other areas, but the range still deteriorated over time. In trying to pinpoint the ultimate cause of the damage, differing viewpoints quickly emerge. Class antagonisms account for one view, as the smaller ranchers blamed Cameron and Greene for the decline in grass coverage. One old homesteader, James Parker, fumed, "These danged big cattlemen gobble up all the land, callin' them grants, then runnin' so much stock on 'em they're destroyin' the range." In part, Parker was criticizing the loss of the open range, which Greene especially had brought to an end. But whereas the fencing of the valley may have ended a way of life on the range (debatable to say the least), the fences no doubt served more to preserve the grassland than to destroy it by allowing for the rotation of grazing between fenced plots. Additionally, the river itself benefitted from the proliferation of wells and stock tanks, which lessened chances of cattle trampling the riparian area along the river channel. Overall, Hadley and Sheridan concluded, "overstocking within the [valley] was probably less severe than in other parts of southern Arizona."[8]

Climate certainly had much to do with the deterioration of the grassland in the valley. The most severe damage occurred during the 1890s drought, when cattle roamed the open range, and competition between stock raisers caused the size of herds to explode. In times of stress the cattle congregated around springs and the river channel. Ranchers remarked on the depressing view of sun-bleached bones strewn along the riverbank, which were thankfully washed away when rains and floods scoured the floodplain. The next drought cycle followed World War I, from about 1918 to 1921. This dry cycle actually lessened the pressure on the valley in the case of the farming homesteads that had proliferated early in the decade, for most of the dryland farmers abandoned their homesteads during the drought. This second drought cycle also occurred after the range had been fenced and the overall numbers of cattle had decreased, lessening the impact of grazing during the dry cycle.[9]

The most extensive farming in the upper basin centered on the traditional areas under cultivation at Santa Cruz, along the stretch of river through the Mascarenas Ranch in Mexico, and near Tubac and Tumacacori in the U.S. In Mexico, a group of about 200 Chinese families occupied land along the river outside of Nogales, Sonora. Some of the families operated their own farms and others worked for the Mascarenas family. The increased cultivation lasted into the 1930s, when rising anti-Chinese sentiment in Mexico forced the families off the land. Vigilante groups drove some of the families off the land at bayonet point. Farming near Tubac and Tumacacori prior to 1930 relied on the fairly constant but meager surface flow downstream from the cienega north of the confluence of Sonoita Creek and the Santa Cruz River, but after 1930 groundwater pumping dominated.[10]

Typical of agriculture in the modern era was the boom and bust cycle of international commodity markets. The experience of the Continental Rubber Company near Canoa exemplifies these new modern realities and the overall difficulties faced by irrigated agriculture in the upper basin and throughout the valley. The Continental Rubber Company bought the northern half of the old Canoa land grant from its owner, L. H. Manning, in 1916, with the intent of exploiting the market for rubber products fostered by World War I. The company set about growing guayule, which was a source of synthetic rubber. Besides stimulating expanded wheat cultivation on the Great Plains and increased cotton production in the South and Southwest, World War I had also sparked production of many other commodities such as guayule. By 1920 the company had over 1000 acres planted in guayule, relying on irrigation from wells and from the Canoa headcut. As with cotton and wheat, however, the post–World War I collapse in rubber prices doomed the project, and the company abandoned the farm. The most profitable operation on the Canoa grant remained stock raising on the southern half of the old grant. In the 1920s, the ranch gained notoriety as a supplier of Arabian horse and prime beef breeding stock.[11]

The early decades of the twentieth century found the twin Nogaleses growing and prospering. In 1910 a railroad line finally arrived directly from Tucson up the Santa Cruz River valley, supplanting the early circuitous route along Sonoita Creek. With the expansion of population and commercial activity, a corresponding development of the water system became imperative. At first the towns' development seemed peripheral to the river's status, since the communities occupied a tributary canyon of the river, but as the cities grew, their reach for water extended to the river itself. In 1911 the City of Nogales, Arizona, bought the private water system and began developing a new well system north of town on the Santa Cruz River. The new well—actually a hand-dug infiltration gallery (much larger than a well, akin to a water-producing cistern)—was situated on the river's bank at the mouth of Proto Canyon where a perennial flow in the river meandered out of Mexico toward the old mission site of Guevavi. The pump house and well were eight miles from the city, and the project included the construction of a pipeline that delivered the water up Proto Canyon. The pumps at the river pushed the water up the canyon to a storage tank placed on a hilltop overlooking the city. From there residents derived their water pressure through gravity flow from the storage tank.[12]

The population in Nogales, Sonora, continued to rely on private wells for most domestic water. A few businesses and residents near the border had access to water from the system on the U.S. side of the border, but the vast majority of residents in Mexico derived water directly from the aquifer. Unlike the aquifer in the middle basin, however, the storage capacity of the aquifer in Nogales Wash was relatively small, extending no more than fifty feet below the surface of the floodplain. This made the water supply vulnerable to changes in climate and variations in rainfall amounts. The shallow aquifer resembled a thin sponge, easily wrung dry. When dry conditions reduced the surface flows and recharge of groundwater, the aquifer quickly became depleted. Such was the case in the dry years of 1921–1923 when wells in the southern end of Nogales Valley, in Mexico, dried up. During times of drought, Sonoran residents bought water from the Arizona company until rains had replenished the shallow aquifer.[13]

In the 1920s, as the towns grew and prospered, the Nogales, Arizona, water department embarked on a construction effort to further develop the water system. The water utility used public funds to improve the pumping capacity at its well by the river, constructed several miles of new water mains, and began chlorinating the city's water to counter a typhoid epidemic in Mexico.[14]

Whether in agriculture or in urban development, the upper basin gave indications of profound changes in the river's circumstances through the first decades in the twentieth century, but the traditional ebb and flow of development continued to determine the circumstances of human society along the river. The draw of water for farms and ranches followed the boom and bust cycle of the 1910s and 1920s, as well as the wet and dry cycles dictated by weather patterns. The new factor of urban development in the basin grew throughout the early decades of the twentieth century. At first the towns merely interrupted the surface flow and subsurface ooze that augmented the Santa Cruz River's flow. But as the communities grew, pumps began taking water directly out of the river, and the modern circumstance of a thirsty industrial and urban society arrived in the upper basin.

Throughout the river valley, water resources and supplies in the early twentieth century still primarily serviced agriculture. The schemes and plans to preserve agriculture in the Tucson Basin continued in the face of declining and often nonexistent surface flows—a continuing trend that would eventually shift the bulk of agricultural production from the middle to lower basin. As groundwater pumping took hold in the twentieth century, surface flows became superfluous to agriculture and the lower basin came to dominate agricultural production in the valley. However, this shift was still decades in the future. In the meantime, Tucson was still the center of much farming activity, and the Allison brothers were in the thick of it. The brothers were instrumental in the effort to expand cultivation in the middle basin, starting with the westside canal, then the eastside canal to the "flowing wells" fields north of town, and eventually in 1900 beginning another canal to the south of town: "we got a right of way from the Indian department at Washington, and dug another ditch, bringing water from the Black Mountain on the Indian Reservation, to land about 14 miles north of the Black Mountain. We cleared the land of heavy mesquite and farmed it for several years."[15]

The Allison brothers had started the process of augmenting the supply of water in their canals with well water. After the Allisons shifted their operation to the south of town, the new owners of the property north of Tucson continued the process of sinking wells. The success of these operations was trumpeted in the newspaper, proclaiming the existence of a "large and permanent supply of artesian water" at a depth of twelve to twenty feet.[16] By 1910 farmers and ranchers along the Rillito River had sunk twenty-five private wells, drawing water from depths ranging from three to thirty-five feet.[17] As more farmers and speculators developed wells in the basin, the newspaper crowed, "This will equal more than one-half the entire water supply of [the] Salt River Valley."[18]

Tucson boosters in 1902 were already recognizing, and were jealous of, the potential benefits to irrigation accruing to Phoenix with the prospect of water from the Salt River Project's dams and distribution systems. No such system ever benefitted farmers in the Tucson Basin. Eventually the Central Arizona Project (CAP) would bring federally subsidized water to Tucson in the 1990s, and in the initial planning of the project, the water was to benefit agriculture in the region. By the time the water finally arrived, however, the vaunted subsidy had melted away in high costs and salinity issues. Federal reclamation proved to be more mirage than panacea for Tucson's farmers.

The lack of a federal connection for Tucson water interests inhibited development of a hydraulic society in the middle basin such as Donald Worster described for portions of the American West in *Rivers of Empire*.

This view from "A" Mountain in 1904 looks northeast over the Santa Cruz floodplain to the city of Tucson in the background. The river meanders past "A" Mountain, with secondary growth of mesquite and cottonwood lining the channel. This view is just downstream from the site of the Tucson Farms crosscut.

Worster's model of federal agencies and their money in league with local power elites holds true more for the Phoenix Basin and the Salt River Project. This was the project that linked federal reclamation, in the form of the Roosevelt Dam, with Phoenix Basin agricultural interests. In the Tucson Basin, national and international capital flowed into the valley to develop water resources, but it remained private, venture-based, elite to be sure, and without federal subsidy or alliance.

Even without federal reclamation, the schemes to augment the water supply for agriculture in the Tucson Basin increased in magnitude and sophistication. Speculators from as far away as Great Britain and Chicago formed the Tucson Farms Company in 1910, planning to develop arable land with enough water to lure midwestern farmers to the basin. The company planned to sell land with "a sufficient water right" to farmers for "two hundred to three hundred dollars per acre."[19] The company started by acquiring land from the Allison brothers and L. H. Manning, and controlled almost 6,000 acres in the middle basin by 1913.[20]

First on the company's agenda was to develop water resources from the Canoa Ranch to the San Xavier Reservation. At Canoa, company workers

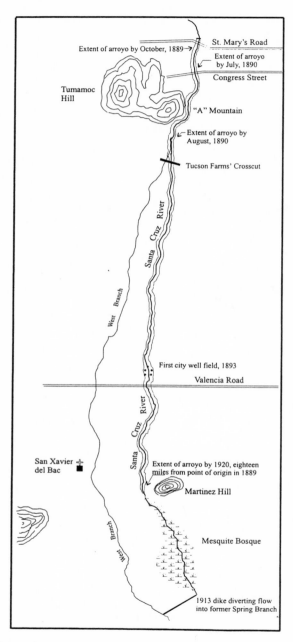

By 1920 the river in the Tucson Basin had undergone dramatic changes. The former spring branch had become the main stem of the river, and the arroyo had eaten its way to Martinez Hill. The ancient cienega south of Martinez Hill had disappeared, but the still-plentiful groundwater supported an extensive mesquite bosque, which was favored by Tucson picnickers and birdwatchers.

dug another trench into the river channel, creating an artificial headcut in an effort to establish more gravity-flow water for irrigation downstream. The company controlled several wells on the San Xavier Reservation, purchased from the Allison brothers, and installed "electrically-operated pumping plants" at those sites, as well as at the two dozen new wells sunk between the reservation and Canoa. The new wells were drilled to depths ranging from 200 to 900 feet.[21]

By far the most grandiose and intricately conceived idea for increasing the water supply was the Tucson Farm's "crosscut system." The company's engineers in conjunction with Bureau of Reclamation surveyors, spent $30,000 drilling test wells and gathering data on the depth to water, depth to impervious rock formations, and the velocity of underground water movements. The company's engineers then conceived of the million-dollar plan to drill nineteen wells in a straight line across the channel of the river, linking each well with a pipe of reinforced concrete almost a mile in total length. The linking conduit was placed five to twelve feet below the level of water. In addition to gravity flow, the system included eight electric-powered pumps, each with a capacity of 1,000 gallons per minute, to augment the supply of water. The system for distributing the water relied on the old Allison canal, but went far beyond the Allison's original conception. As the chief engineer described:

> "The distributing system consists of a reinforced concrete pipe line 48 inches in diameter, 1500 feet in length forming the outlet from the recovery system, a 48 inch concrete siphon under the Santa Cruz river, about 7 miles of ordinary earth canal, 12,650 feet (2.4 miles) of which are lined with concrete, and 21 miles of laterals of which about 3000 feet (0.57 miles) are lined with concrete. In addition there are numerous reinforced concrete drops, provided with steel measuring weir, steel flumes, 1200 feet of 24-inch concrete siphons under the Santa Cruz and Rillito Rivers, one earth and one timber dam across the former river, a large number of check and division boxes, lateral head-gates, drainage structures, etc., about 200 in number and common to most irrigation projects."

The nineteen wells ranged in depth from 45 to 150 feet below the surface of the ground, and the construction, so much of it taking place below the water table, proved understandably taxing and complicated.[22] By 1915—a very wet year—the system delivered almost 30 million gallons of water a

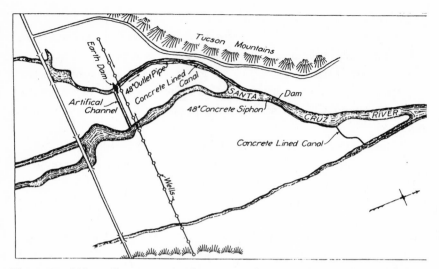

The most ambitious effort to augment the water supply using wells in the middle basin was the Tucson Farms Company crosscut. It linked nineteen wells that extended for a mile across the floodplain. Eighteen pumps created an initial output of thirty million gallons of groundwater a day.

day. But even this was not enough for the land-speculation scheme. As one observer of the project later remarked, "The trouble with building the dam thing was too much water, and then it was never a success because they never had enough."[23]

One aspect of the Tucson Farms Company's plan was to reconstruct part of the Allison brothers' old system, which the company's engineers found to be "very crudely built. . . . For the most part the gradient was very low and poorly equalized while the location of the ditch through a part of the city inhabited largely by Mexicans, running as it did through alleys, back yards, and under houses with poorly defined rights of way, made the re-location a rather troublesome matter." The old canal had been unlined and the engineers calculated that it lost five percent of its water per mile, not only through seepage but through evaporation and the "transpiration of moisture by the rank growth of vegetation which in that warm climate grows profusely and soon fills the canal section." The engineer recounted how the old canal had required yearly cleaning—part of the old acequia regimen dating back centuries—and trumpeted the virtues of the new, relatively maintenance free, concrete-lined canal.[24]

The ups and downs in the cotton market affected farmers in the middle

The Tucson Farms Company hoped to lure midwestern farmers to the Tucson Basin with the promise of cheap land with ample water for agriculture. A major selling point for the company was the use of concrete-lined canals to deliver water from the wells and crosscut to the fields. The initial seven miles of the distribution system included about two and a half miles of concrete-lined canals.

basin just as fluctuations in the guayule market had affected production in the upper basin, and the Post Project serves as a prime example. Edwin R. Post, one of the heirs to the Post cereal fortune, sought to emulate the Tucson Farms Company by luring farmers to the basin by promising irrigation water for cotton production. Post bought land between the Rillito confluence and Marana and drilled ten new wells in the floodplain during the cotton boom in World War I. To power the well pumps, Post installed a "1,500 kilowatt steam turbine generator," but when cotton prices dropped after the war, many of the immigrant farmers went bankrupt. Eventually the land changed hands, becoming first the Pima Farms Company, and then Carder Farms. With the rebound in cotton prices in the late 1920s, cotton production increased, but by then sharecroppers and leaseholders conducted the farming. Similarly, the Tucson Farms Company land changed hands in the 1920s. The Flowing Wells Irrigation District formed in 1922, and took over control of the Tucson Farms Company crosscut. Also in the 1920s, Midvale Farms acquired the company's land between the San Xavier Reservation and the city.[25]

This view south from "A" Mountain in 1927 shows agricultural fields and the river still meandering toward Tucson. In the 1930s the channel was straightened by dumping old car frames into the river at the point of the meanders. In the distance is Martinez Hill, with the Santa Rita Mountains in the far background.

One of the difficulties faced by agriculture in the middle basin was the growth of the arroyo, which between 1910 and 1920 had extended three miles into the Indian reservation, destroying 150 acres of arable land in the process. The arroyo now extended eighteen miles from its point of origin at the head of the Hughes ditch near St. Mary's Road. The destruction of Indian land caused the federal government to launch a study of the problem, and the government scientists dispatched to do so, Olberg and Schanck, recommended the construction of a dike between the two stream channels south of Martinez Hill so as to direct future floodwaters into a manageable single channel that basically followed the eastern spring branch of the river. "This would cause the flood water to enter the Tucson channel at the Sahuarita Butte [Martinez Hill], where the erosive action would do but little damage to the tillable land." One problem with the plan was litigation brought by the Tucson Farms Company. The company had land near the proposed dike and had constructed "an expensive pumping plant" next to the main channel precisely where the dike would funnel the future floods into the single channel. On the other hand, an advantage of the dike would be the creation of "a flood storage reservoir" that could catch and hold up to 350 acre-feet of water. Olberg and Schanck proposed the dike in

1909, and the channel was constructed in 1913. The existing channel of the river into Tucson was in place with the construction of this dike.[26]

In the early decades of the twentieth century, civic leaders in Tucson considered the economic health of the community to be tied to traditional economic pursuits. Starting in the 1920s, boosters described these sources of prosperity in Arizona as the "Three C's," cattle, cotton, and copper, as indicated in the Chamber of Commerce logo proclaiming mining, agriculture and ranching as key to Tucson's economic development. Tourism was increasing in its visibility and significance to the city in the 1920s, eventually adding a fourth "C," climate, to the triumvirate. But to many city leaders agriculture and mining remained the key to prosperity in Tucson. A newspaper editorial reflected this view in June 1923, lamenting that "agriculture projects [were] now deterred by lack of data on valley water resources." Booster organizations coalesced behind the effort in 1923 to get a scientific study of water resources in the valley, even though the city council refused to appropriate any funds to help pay for the study. Acquisition of data concerning water resources became a mantra at times, reflecting a lingering progressive assumption that expert knowledge could solve all problems. As the district engineer for the U.S. Geological Survey (USGS), Roger C. Rice, declared, "Agriculture in the Santa Cruz valley would receive an inestimable stimulus from the preparation of a thorough and accurate survey of the water resources as a basis for future projects." The possibility that an accurate survey might place limits on future projects did not often intrude into the engineer's, or the boosters', rhetoric. Whereas the city council in 1923 balked at financing a water resources survey, subsequent councils willingly ponied up funds for such purposes.[27]

Although agriculture remained dominant, in the first decades of the new century clear signs emerged of the increasing importance of urban areas in both the utilization and management of water resources. The development of municipal water systems serves as a clear indication of the rising urban demand for water in the river valley. The municipal system in Nogales placed only a modest demand on the river's water supply well into the twentieth century, but in the middle basin, Tucson's thirst had already made noticeable inroads into the aquifer's supply.

In 1900, during the era of progressive reforms of public utilities, the City of Tucson bought the privately owned Tucson Water Company for $109,000. At the time of the purchase, the water company delivered water to 625 connections, had constructed 46,975 feet of water mains, maintained 62 fire hydrants, and collected $18,000 in gross revenue, netting $7,440. For the population of about 7,500 in Tucson in 1900, the water company pumped about two million gallons of water a day. That translates into a whopping 266 gallons per person per day (compared to Tucson's 1922 rate of 176 gallons per person per day). The primary reason for the high rate of water use in 1900 was the continuing practice of the water company to use some of its water for the irrigation of agricultural land. The operating costs of the water system were $10,560 a year.[28]

Of primary concern for Tucsonans and the Tucson Water Company was the source of the city's water. First there was the river, entrenched through a broad floodplain, inexorably sinking beneath the sand in the early decades of the twentieth century. Then came the shallow wells in and near the river's bed, followed by even deeper wells scattered throughout the basin. Soon city leaders and boosters began casting covetous glances at nearby watersheds and drainages contemplating schemes for capturing their potential sources of water. Through all the plans and developments, however, Tucson remained dependent on groundwater pumped from the aquifer of the Santa Cruz River and its tributaries, and would do so until the Central Arizona Project finally delivered Colorado River water to Tucson in the 1990s.

Over the early decades of the twentieth century the numbers of wells increased, and the city's reliance on groundwater pumping became a matter of absolute dependence, or in a manner of speaking, an addiction. From 1893 to 1922, the city instituted seven water-related bond issues and revenue measures. The city passed an ordinance in 1916 "fixing the rates for water service through meters." In 1922 the water company doubled its supply of water with the addition of three wells east of the University of Arizona and a "northside plant" with four new wells. By the time the 1920s ended, the water system included scores of wells in two "zones," one eighty feet higher than the other. The growing system required "booster stations" to provide water pressure to customers far removed from the river and floodplain. The urban geography of Tucson was already a fact of life for water company engineers. The pattern of Tucson's expansion had been established to the east, ever further from the river and higher in elevation from the original floodplain.[29]

As the water system expanded in girth and sophistication, city leaders and bureaucrats came to realize the problematic nature of a reliance on groundwater, and the search for additional sources of water became a common topic of discussion at service club meetings and city council planning sessions. In February 1923, the city engineer, G. H. Atchley, discussed a plan to use water from the Pantano Basin twenty-two miles to the east. Later that year, the leading hydraulic engineer at the University of Arizona, G.E.P. Smith, encouraged city leaders to build a dam in Sabino Canyon: "the city should build the Sabino Canyon dam and reservoir not to replace the present water supply, but to supplement it. It could be built solidly and not hurriedly and would be insurance against a water famine in the city in the event of something going wrong with the present supply. To be a beautiful city Tucson must have lots of water. It is an important factor in the development of the community."[30] The plan was a tough sell given the recent doubling of the city's water reserves, and the city never completed a reservoir in Sabino Canyon, much to the pleasure of future picnickers, cyclists, and hikers, albeit to the chagrin of valley boaters and fishermen. Smith's suggestion nevertheless points to the more-or-less ever-present concern of "water famine" in the Old Pueblo, and the notion that prosperity in the community required "lots of water."

Engineer Atchley and hydrologist Smith represent the blossoming progressive faith that technology and science could solve the problem of the region's scarcity of water. Smith had produced in 1910 the first of many studies of groundwater in the basin through his position at the University of Arizona and the Agricultural Experiment Station. The 1910 study is interesting both for its optimistic appraisal of the prospects for expanded water supplies in the basin, and for Smith's observations about the status of the Santa Cruz River. Groundwater pumping was already problematic to Smith in 1910, and only his progressive faith in dams and reservoirs overcame his logically founded concern: "It may be questioned whether it is right to draw more heavily from the groundwater supply than is reasonably certain of renewal each year. Yet this question is not raised in the case of artificial reservoirs." Smith's other observation of interest to this study was his appraisal of the Santa Cruz River in 1910 as "ever a dwindling stream." Smith had centered his study of water supplies in the basin on the Rillito because he judged its resources to be potentially more bountiful than that of the Santa Cruz: "Only by courtesy can the Rillito be said to empty into the Santa Cruz, for the latter is but a brook compared with the raging floods of the former. At the junction of the two rivers the width of the Santa Cruz is 60 feet while the width of the Rillito (tributary) is 300 feet, and the river beyond has the appearance of being the continuation of the tributary."[31] Only one year previously, Olberg and Schanck had referred to the Santa Cruz as a "so-called" river. Clearly, the old stream had fallen on hard times.

The effort to develop a water system capable of delivering peak quantities of water to the City of Tucson accompanied the effort to locate and exploit additional water sources. An early indication of these modern realities in the Tucson Basin occurs in the vocal and urgent demands of Tucson residents for consistent and reliable water service. These political developments indicate a definite wagging of the urban tail. A clear history of shortages and interrupted service faced every political leader in Tucson in the early twentieth century. For example, in 1903 the city water company appealed to "the fairminded citizens of Tucson" to curb their watering of lawns and gardens during peak usage hours.

> Notice to Water Consumers;
> Owing to the increased consumption of water so far in excess of the present means of supply and the decrease in the underground flow to the now existing wells, it becomes necessary to curb the sprinkling and irrigating of lawns and trees, until the installation of the new pump ordered by the city, and the completion of the new well now under course of construction, to the following hours, viz.—
> Sprinkling and irrigating allowed only between the hours of 5 a.m. and 8 a.m. and between 5 p.m. and 8 p.m.
> Trusting that the fairminded citizens of Tucson will bear with us in this proposition, Yours respectfully, Philip Contezen, WM Reid, F. J. Villaescusa. Water Committee.[32]

This 1903 announcement was not the first incidence of a request for voluntary limiting of water usage, although it did mark a certain stiffening in the tone of social control. In 1893 the newspaper had passed along a tentatively worded request for similar purposes: "It has been suggested that some plan should be adopted by which all irrigation of gardens, lawns and trees in Tucson should be done from six o'clock in the evening to four o'clock in the morning. This plan would be much better for the gardens and lawns and would make the water supply ample during the dry season of the year."[33]

Outside the realm of request or cajoling, there was also the defacto circumstances of limited availability when the water pipes simply ran dry.

In June 1893, customers complained that no water flowed through the pipes until after eight or nine o'clock in the morning, and the newspaper warned, "The water supply at the water works is running so low that patrons will do well not to be extravagant in using from their hydrants until rains come to replenish the supply. A fire to happen while the pressure is as light as it is some days would be a serious affair."[34]

Whereas issues of supply rested finally on physical limitations and realities in the semiarid bioregion, decisions about the water system resided more firmly within the political realm. This has been more-or-less a constant in urban circumstances, at least since industrial cities began to cope with domestic luxuries such as flush toilets and bathtubs in the mid-nineteenth century. As more and more homes and businesses adopted the modern conveniences, cities had to construct and maintain huge water works. For example, Boston faced a water crisis in the 1840s, as the new innovations proliferated in hotels and middle-class homes. Demand for water outstripped supply, not in the sense of an ultimate water famine, but in an antiquated system's inability to deliver a volume sufficient to meet demand. A major overhaul of the system in the early 1840s established a capacity of 7.5 million gallons a day, but daily water usage in Boston had increased to 8.3 million gallons a day in 1843. In other words, the new system was rendered obsolete soon after it came on line. By 1860, Boston's water usage had increased to 17 million gallons a day, a per capita use of 97 gallons of water a day. Police walking the beat at night listened for running water, a circumstance that would result in a rap on the door and stern admonishment, a relatively mild form of social control, but indicative nonetheless of intensifying water politics.[35]

A scandal in the Tucson water department in 1921 serves as another sort of marker of the Old Pueblo's maturing water politics. By that time Tucson's water system had grown to such a size that work crews toiled more-or-less constantly on construction and maintenance projects. What might seem an unrelated matter of petty civic politics had at its heart the river's aquifer. A bureaucracy had grown complete with sinecures and fiefdoms around the water supply and delivery system—at least the partial manifestation of a hydraulic society in Tucson. In September 1921, the county attorney, George H. Darnell, accused the superintendent of the water department, Percy C. Smith, and the foreman of the "pipe gang," Ramon Montez, with filing false time cards and collecting kickbacks from workers on the city's crews. In the shake-up that followed, the city council appointed G. H. Atchley, the city's chief engineer, to head the water department. The chief engineer later reported that he had found the water department in a sad state of disrepair, and set upon a one-year program of repair, maintenance, and development. After that program, Atchley reported in 1923 the aforementioned doubling of the city's water supply.[36]

Chief engineer Atchley gave a statistical snapshot of Tucson's water usage in his 1923 report. The city's wells could provide 14 million gallons a day, which the engineer calculated was sufficient to supply a population of 61,000. This calculation was based on daily water use by Tucsonans in 1922 that averaged 176 gallons per person per day, and a peak demand during June 1922 of 227 gallons per person per day. Tucson's population at the time was about 25,000.[37] Whereas the average use of water per person per day was down from its 1900 high of 266 gallons (due to the end of the water company's agricultural pursuits), the high peak rates in 1922 were largely due to the practice of flood irrigation of lawns and gardens during the summer. Common in Tucson were landscaped lawns surrounded by berms of about a foot in height. The grass, shrubs and trees in the lawn were treated to watering by filling up the small ponds created by the berms. The University of Arizona, maintaining its own water system and wells independent from the city system, also practiced flood irrigation of the campus grounds, and the mall areas on the older parts of the campus still show the old berms and irrigation mechanisms.

With a water supply far in excess of the city's needs, consensus behind bond issues and development plans wavered. The city council refused to allocate funds to support a water resources survey, and also in 1923, a proposed $700,000 bond election for water projects ran into resistance. An editorial in the *Arizona Daily Star* urged caution in the city's acquisition of additional debt because $750,000 in recent bonds passed for the construction of the new high school had become too expensive. The newspaper pointed to the current "deflation period" as an inopportune time for acquiring debt. In the end, the city council delayed, and construction of new water mains was slowed. But such temperance never lasted for long. Either drought conditions heightening the fear of water famine, or sprawl lowering the water pressure in the city's system, created the political support necessary for another round of bond elections.[38]

The lower basin of the river came to life with irrigated agriculture at last in the twentieth century. Long dormant in terms of economic development, save the tenuous thread of activity along the stage road and rail line, the dry flats required industrial technology to provide the means of tapping the vast underground supply of water. Once the means were at hand, only the will and desire to utilize the technology remained to surface, and it was not long in coming.

An early scheme for the lower basin was part of the overall plan for the valley by the Tucson Farms Company. The Santa Cruz Reservoir Project intended to catch runoff from the river and its tributary washes far downstream in the lower basin, near Picacho Peak in Pinal County. The resulting reservoir would provide irrigation water for more transplanted farmers. The engineer in charge of the project, P. E. Fuller, had noted the result of the Hughes ditch in the Tucson Basin, and fully expected the excavations in the river channel far downstream to have the same effect, widening and gouging the channel to the reservoir at the next flood.[39] The investors, including L. H. Manning, poured $10 million into the construction of levees, canals, dams and catch basins designed to produce a reservoir holding over 97 billion gallons of water (300,000 acre-feet). The most ambitious part of the plan was a canal thirteen miles long carrying floodwater from the river channel to the reservoir. In 1909 the land company planted crops in the bed of the reservoir and diverted "several thousand acre-feet of water" from the flooding Santa Cruz into the reservoir. Ultimately, however, this effort to stimulate development of the lower basin failed. Schemes to cultivate the lower basin with surface flows, no matter how intricately conceived, were doomed, primarily due to the high evaporation rates of the desert terrain.[40]

The Santa Cruz Reservoir Project also serves as a testament to the stubborn persistence of the archaic perspective of the river. The belief that surface water carries an inherently desirable, even mythical, nature lingers to this day, even among otherwise "modern" residents of the valley. As cases in point, proposals recently surfaced in Pima and Pinal County to construct two lakes relying on constant infusions of groundwater or CAP water (to be discussed in chapter 12).

Only the mining of underground water supplies accomplished the goal of establishing agriculture in the lower basin. Farm companies at Carder, Marana, and Red Rock (formerly Cestui) profited from the cotton boom during World War I, and then struggled to survive the crash in cotton prices in the 1920s. Despite the ups and downs in commodity markets, the irrigated farms in the lower basin survived to replace the upper basin as the center of agriculture in the river valley. By the mid-1930s farmers used groundwater to cultivate over 98,000 acres of land from the confluence of the Rillito and Santa Cruz to the confluence of the Santa Cruz and Gila River. In comparison, south of the Rillito to the Mexican border, slightly more than 22,000 acres remained under irrigated cultivation.[41]

Completely irrigated agriculture freed farmers from their dependence on rain, or at least, such was the hope in the early 1900s. Thwarting this ambition were the continuing drought conditions of the 1890s and into the twentieth century. In response to the declining surface flows in the middle basin, wells proliferated, as has been noted in this and the previous chapter. The increasing number of wells had resulted in the newspaper's boosterish proclamation comparing the Tucson Basin's water supply to the Salt River Valley's. But as could be expected, the severe drought years of the early 1900s tempered optimism. Water shortages returned as output from the wells declined. One explanation for the apparent declining water table was the steady erosion of the river channel, which observers noted was acting like a giant drainage canal, siphoning off the underground water on each side of the channel.[42]

Climate continued to affect the river and human society in the valley, even as industrial society endeavored to render climate inconsequential. Drought cycles run their course, even in the semiarid Southwest, and such was the case as the 1900s progressed. Rain amounts increased through the decade, until the winter of 1914–1915 brought record amounts of rainfall. The rain in December 1914 created the highest recorded flow past Congress Street to that date, recorded on December 23 at almost 15,000 cubic feet per second—that is to say, one acre-foot of water passed Congress Street every 2.9 seconds (remember Culver's 1884 report that during low-flow periods, the Santa Cruz could reliably provide one acre-foot every 45 minutes). The total flow past Tucson that month was almost two million acre-feet. The floods that winter destroyed bridges, pasturage, and houses; caused $10,000 damage to the Tucson Farms crosscut; ruined city wells; destroyed Southern Pacific Railroad tracks in several locations; and in general wreaked havoc throughout the valley.[43]

One result of the 1914–1915 floods was the arroyoization of the channel in the lower basin near the Santa Cruz Reservoir. The flood gouged a channel along the ditch just as the engineer Fuller had foreseen. Betancourt acknowledged that the ditch, along with the Hughes ditch in the Tucson Basin, contributed greatly to the creation of the discontinuous arroyo and entrenching of the river channel.[44]

The Tucson flood of December 1914 set the record for the greatest flow past Congress Street to that date. Similar flows did not occur again until the 1960s, and a new benchmark for floods did not occur until the record flows of 1977 and 1983. This photograph shows Tucsonans watching the spectacle as another section of the riverbank splashes into the channel. Each successive flood deepened and widened the arroyo in a process that began in 1889 with the Hughes headcut.

Once again, weather patterns turned and dry conditions returned later in the decade, continuing into the early 1920s. Water shortages plagued the cities periodically, most often in relation to these dry seasons and droughts, though due as well to limitations in the water departments' systems of wells and mains. Restrictions on water usage in the summers in Tucson had come to be expected by the early 1920s. The same dry conditions affected the aquifer in the Nogales Valley.[45]

Droughts interspersed with rainfall peaks continued the process of erosion as human machinations in the river channel increased, further accelerating the erosion process. One of the last sections of channel to form was along the old dirt road from Tucson to Nogales. The creeping arroyo engulfed the old wagon road "on the westside of the valley south of San Xavier," eventually adding this channel to the main flow of the river. With that, and the earlier construction of the dyke on the reservation, the river's point of entrance and primary course through the middle basin was established.[46] By 1990 the channel had grown in places to twenty-four feet deep and over 300 yards wide.[47]

Charles Lindbergh flew the Spirit of St. Louis into Tucson in 1927 during his national tour following his famous transatlantic flight.[48] Circling Tucson's Davis-Monthan Air Field prior to landing, Lindbergh would have viewed a scene with many markers of change accomplished, and many indications of change yet to come. The city was already sprawling to the east with residential grids and commercial thoroughfares crisscrossing the desert. Agricultural fields still occupied much of the old floodplain, but if Lindbergh had coaxed the Spirit of St. Louis up to 10,000 feet or so, he could have seen many more fields spreading across the flats to the northwest. More fields marked the river channel to the south, and if Lindbergh had turned the nose of his plane in that direction, he would have viewed an upper basin sprouting with agriculture along the river, cattle on the hillsides, and a hint of urban development in the distant canyons of Nogales. From high above the San Rafael Valley, faint gridlines may have appeared in the grasslands below as fencing regulated where cattle grazed. Some patches would have appeared thick with grass, while others recently grazed would have shown a dusty brown.

Lindbergh was promoting aviation—an icon of modernism—on his national tour in 1927. A no more fitting spokesman could the fledgling industry have found—the pilot had become a national hero of unprecedented stature. All eyes looked up as cheering crowds welcomed the aviator and his famous airplane to the Old Pueblo. The river had little to do with Lindbergh's appearance and no doubt few thoughts during his brief stay turned toward the river, its aquifer, or the stream's changing circumstances. That collective moment of disassociation from the river gives a hint of the later postmodern perception held by the valley dwellers, but in the late 1920s no degree of distraction from basic water realities could last for long. Droughts and floods, periodic system failures and shortfalls, and arroyoization had become recurring features of life in the city. In the countryside, modern agriculture required wells and pumps and was no longer tied to the diminishing surface flow.

Lost Balance and "The Truth About Water in Tucson"

1930–1960

THE DEPRESSION YEARS OF THE 1930S directly affected the river in many ways, but not in a single, clearly marked direction. In places such as the Carder Farms area north of Tucson, declining agricultural production resulted in reduced pumping of groundwater. In response, the water table rose as much as sixteen feet during the early 1930s. On other stretches of the river agriculture persisted and urban demand expanded, creating increasing draws on the river's underground supply.

The decade of the 1940s began with signs of rising prosperity and optimism throughout the valley after years of depression and pinched circumstance. With talk of a world war on the horizon and with the bubbling enthusiasm of the boosters and businessmen throughout the valley, it was easy to ignore the changing circumstances of the river, yet it remained the river's task to bear the weight of the valley's expanding development. The assertion of lost balance between withdrawals and recharge in 1940 stems largely from the recognition of greatly increasing drafts—overdrafts—from the groundwater supply starting in 1940 and further increasing through the 1950s. Although the water table had begun dropping prior to the turn of the century in the Tucson Basin, the decline was slight and relatively easy to dismiss. The declining water table first became noticeable, then undeniable, and finally startling during the war years and postwar boom.

But all seemed rosy in 1940. Agriculture in the valley was enjoying rising commodity prices as world conflagration placed U.S. producers on the positive side of the supply and demand equation. Marketplace developments thus spurred agricultural production, especially in the lower basin, and ranching and farming flourished in the upper basin as well. In the middle basin, agriculture saw a decline as the City of Tucson grew and prospered.

As the middle basin became more and more dominated by Tucson's growth and urbanization, water politics in the entire valley came under the influence of the shimmering metropolis in the desert. Reflecting the increasing influence of Tucson over water resources in the valley, the remaining chapters in this narrative shift focus to the story of Tucson's water politics—a water politics more rancorous, divisive, and intractable than any other in the valley. After World War II, Tucson *became* the middle basin, pushing agriculture out, spreading over the former floodplain, and soon sprawling over the entire basin.

In some ways, developments in the Tucson Basin were mirrored by those in the upper basin, primarily in the urban development in Nogales Wash. But in the realm of agriculture, the upper basin persisted in the effort to maintain production while the middle basin saw agricultural production drastically curtailed. Ranching in the upper basin continued in the manner pioneered by early efforts to preserve the grassland in the San Rafael Valley. By the 1950s probably the only stretches of the river that bore any resemblance to their prehuman or even preindustrial appearance were found in the high valley in the upper basin. The mountains still glimmered in the distance, grass whispering and trees rustling as a meager but steady stream passed through the valley in service to farms and ranches, with domestic uses so slight as to be negligible. As the river made its U-turn in Mexico, the passage through the Santa Barbara Ranch marked a continuity with the past decades: cattle grazing on the range and persistent farming along the river bottom. But the twin cities of Nogales in Sonora and Arizona were growing, and the community of Tubac farther downstream continued to expand in population as well.

In an effort to maintain and even expand agricultural production during the Great Depression, farmers in the vicinity of Tubac and Tumacacori removed thousands of cottonwoods that lined the stream banks. The cottonwoods, deemed thirsty by the pastoralists, were killed by ringing; that is, cutting the bark around the circumference of the tree, thus cutting off the flow of nutrients up from the roots. Once the trees were dead, they were set afire. Douglas Cumming, long-time rancher and observer of the river in the vicinity, measured one of the dying trees to be thirty-five feet in circumference. The burning trees must have formed magnificent pyres—a spectacular way for the giant cottonwoods to go, at least compared to their neighboring willows and cattails. The smaller trees and shrubs simply

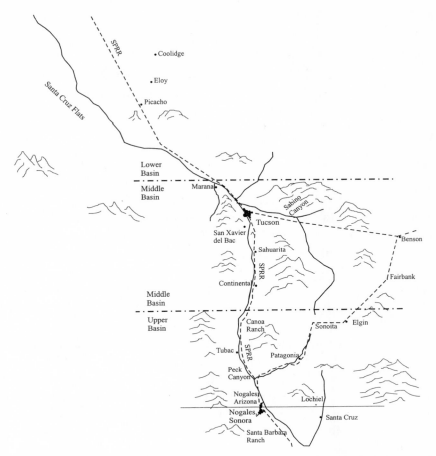

By the 1940s, the proliferation of groundwater pumping in the lower basin had spawned the development of new communities downstream from Tucson. Not since prehistoric times had human societies lived in the dry reaches of the lower basin. Elsewhere in the valley, returning prosperity during World War II spawned Tucson's growth, even as agriculture fled the middle basin to the north and south.

wilted and withered as the water table declined with steady pumping. The thick bosques of undergrowth, such as Charles Shuchard had sketched in 1854, disappeared in the 1930s. On the hillsides, however, mesquite began spreading. Cumming attributed the expansion of mesquite onto the hillsides formerly covered with grass as being primarily due to overgrazing. The cattle herds had grown as predators like the Mexican wolf had been exterminated, and depression-era market pressures pushed ranchers into maximum production efforts. The depression ultimately hit the farming community at Tubac hard. Despite the farmers' concerted efforts to maintain and expand production, by 1940 only fifteen families remained in the vicinity of the old presidio.[1]

Ranching in the upper basin suffered through another drought episode in the 1930s—the same dry cycle that caused the Dust Bowl in the southern Plains—but this drought found ranchers in the upper basin better able to cope with the tight circumstances. "Offtake strategies" (ways of getting cattle to market or to other pasturage) had improved, primarily through the ability to remove cattle from the range by truck. Windmills and stock tanks, as well as developed springs, kept cattle from concentrating along the riverbed and trampling the riparian area. The size of the herds was also reduced, or at least regulated, after the creation of forest reserves in 1891, and the Taylor Grazing Act in 1935, which gave the federal government the ability (if not the will) to govern the numbers of grazing livestock on the range. Additionally, during the 1930s, depression economics drove many of the small ranchers off the land. Even with these mitigating circumstances, however, the mid-1930s drought "did considerable damage" to the grasslands of the San Rafael Valley.[2]

The dry cycle eased in the 1940s, but another severe drought hit ranchers in the 1950s. In the face of these difficult circumstances, the Heady and Ashburn Ranch in the San Rafael Valley accomplished a successful range restoration program that became widely known throughout the state. The ranchers had found the land damaged by overgrazing, which allowed erosion and the proliferation of weeds over a wide area. To restore the range, Heady and Ashburn employed methods from the most prosaic to technocratic: pulling weeds by hand and building simple "spreader dams," catch basins, and other water-distribution systems. In the words of an admiring reporter for the state's ranching journal: "The native grasses gradually came back, until today, even in the driest year on record [1953], there is an abundance of feed despite normal grazing . . . the eroded gullies have grassed and sodded up and it seems obvious that one of Tom Heady's objectives has been accomplished, namely, that whatever rain does fall on the place stays there and seeps into the ground because the grass holds it there." The reporter went on to instruct his readers of the key to Heady and Ashburn's success. The ranchers had consistently avoided the temptation during wet years to increase their herds beyond their conservative level of thirty head per section. The relatively small numbers of grazing livestock allowed some grass to seed out, perpetuating the grass growth even through dry years. Whether this practice could or would be widely copied was of course problematic, but Heady and Ashburn were not alone in expressing that hope and desire.[3]

As previously mentioned, an effect of the continued ranching activity in the upper basin as observed by Douglas Cumming was the spread of mesquite onto the mesas and slopes overlooking the river. Cumming noted that overgrazing had thinned out the grass so much that brush fires rarely swept over the hills. These fires had served to kill off mesquite seedlings in the past. Now the seedlings took root and brushy mesquite began to dominate the grasslands. Cumming remained optimistic, however, that by fencing the range and by using more enlightened grazing practices the old grasslands would be restored to something of their former glory.[4]

The further commercialization of the ranching industry in the upper basin also affected declining water tables. To illustrate this trend, a ranching operation had developed in the upper basin that was indicative of the process begun by Cameron's innovations in the late-nineteenth century. A feedlot with a capacity of 3,000 head was operating at Agua Linda near the old Canoa land grant in the 1950s. The owner, Carlos Ronstadt, also cultivated 450 acres on the ranch, which produced all the feed for the cattle on the lot, except for molasses. The farm was irrigated by seven wells at a depth of about forty feet, producing as much as 2,000 gallons a minute.[5]

Other developments maintained pressure on the river. The Mascarenas family continued to graze livestock on the old Santa Barbara Ranch, albeit in a less monolithic fashion. The heirs to the land had divided the holdings into four equal shares in 1949, but the aggregate effect of this land division was minimal at first. The heirs continued to run cattle on the hills and raise crops along the river. Eventually the federal government in Mexico enacted restrictive policies to limit floodplain farming on the Mascarenas property, but these policies remained sixteen years in the future as the heirs divided the property in 1949.[6]

The urban centers in the upper basin continued to grow through the depression and war years. Nogales, Sonora, continued to outpace its U.S. neighbor, maintaining its three-to-one advantage in population through the twentieth century. Growth had taken place in spite of the periodic floods in the valley, which were an increasing concern in the 1930s as property and lives became more vulnerable. The shallow aquifer in the valley quickly became saturated with monsoon and winter rains, especially as commercial development paved over more of the floodplain. Houses also spread up the hillsides, and the narrow canyon became ever more susceptible to flooding. A series of "major floods" occurred in 1905, 1909, 1914, 1915, 1926, and 1930. In particular, the 1930 floods drew the attention of later New Deal officials, having caused five deaths and much property damage. Senator Carl Hayden spearheaded an effort to plan and finance a flood control project for the border communities that was constructed through the 1930s and into the 1940s. Once completed, Nogales, Arizona, sat atop three miles of "covered channel," with another mile and one-half of "lined channel" extending beyond the urbanized area to the north.[7]

On a much smaller scale, but significant nonetheless, Tubac continued to grow in population, though no longer as an agricultural community. The establishment of an art school in Tubac in 1948 had started the process of converting the old presidio into an artists' community. The 1950s also saw the creation of Tubac Country Club along the river. Tubac thus began developing along the twin paths of artists' colony and retirement and resort community. By far, the largest draw on water in Tubac went to service the golf course. The country club relied primarily on groundwater to create the lush fairways and greens. Healthy grass was just as serious a concern for the resort managers as it was for Tom Heady in the San Rafael Valley, but the grass in Tubac would never be grazed by four-legged livestock.[8]

In a major development affecting the river in the upper basin in the 1950s, the discharge of treated effluent from the Nogales sewage treatment plant placed hundreds of thousands—and eventually millions—of gallons of water a day into the river channel, returning a perennial flow to the river that had ceased in the 1930s. The treatment of wastewater from both cities of Nogales had begun in 1951 at the plant constructed one and one-half miles north of the border. The initial flows into the facility were 700,000 gallons a day from Nogales, Arizona, and 100,000 gallons a day from Sonora. Sewage volumes on both sides of the border increased in the 1950s, and the Sonoran percentage of wastewater entering the plant increased from 15 percent in 1951 to 40 percent in 1957. As the flow into the plant reached its maximum capacity of 1.6 million gallons a day, both cities faced moratoriums on new connections to the sewage system. Mexico's volume of sewage had increased even though about half of the homes in Nogales, Sonora, were without connections to the city's drainage system. Thus even with increasing flows into the treatment facility, raw sewage continued to flow down Nogales Wash toward the Santa Cruz River.[9]

Even though the treatment plant remained in a race with the growth of the two cities, the discharge from the plant renewed the perennial flow in the river from Nogales Wash to beyond Tubac. At times, such as during the dry months before the summer monsoon began in July, and during dry years when yearly precipitation totals fell below the norm, the effluent in the river came to account for about 80 percent of the surface flow.[10]

Downstream in the lower basin the growth of irrigated agriculture from 1930 through the 1950s was truly phenomenal. The river channel had disappeared in a maze of irrigation canals and desert washes, but the river's aquifer remained a constant source of water for the new irrigation districts in northern Pima and Pinal counties: Avra-Marana, Eloy, Casa Grande-Florence-Sacaton, and Maricopa-Stanfield.

An early sign of the increasing prominence of agriculture downstream from Tucson came in 1936–1937 when the Soil Conservation Service (SCS) studied flood control problems on the Santa Cruz River. Of primary concern to the SCS was the development of the arroyo through the Tucson Basin, which had "shifted hazards from flood inundation downstream." The investigation by the SCS also looked at problems on the Papago Reservation, where farmers were shifting from farming to woodcutting due to erosion of the fields. The fear was that chopping down trees near the river would increase the likelihood of damaging erosion downstream. Since the channel through Tucson was transferring the hazard to the lower basin, the SCS encouraged efforts to return the Papagos to agriculture.[11]

The increase in groundwater pumping due to World War II's influence on cotton production, as well as on other agricultural commodities, was nothing short of amazing. In 1941 at the start of the war, farmers had pumped 344,000 acre-feet from the aquifer; by 1946 that amount would almost double to 672,000 acre-feet annually. Pumping surpassed one million acre-feet in 1949, and the total for 1951 stood at 1,110,000 acre-feet. The water table had declined an average thirty-seven feet from 1942 to 1952 as a result of the removal of 7.6 million acre-feet of water. Hydrologist R. L. Cushman estimated in 1952 that by the time ten million acre-feet were removed from the basin, the water table would have dropped an average of fifty feet. It took over 1,550 wells to accomplish such a huge overdraft.[12]

A visible result of the groundwater overdrafts in the lower basin was the subsidence of the land surface above the aquifer. Although the land above the declining water table began to sink at a slow and steady rate that was difficult to observe, occasionally the rate of subsidence was uneven and sudden, resulting in "differential subsidence" that caused abrupt shifts in the land surface. One such occurrence, the "El Grande" fissure near Picacho in the lower basin, eventually grew to hundreds of feet wide and ten miles long. In most areas, however, the land in the valley subsided gradually. For example, in the lower basin near Eloy, the land surface dropped about fifteen feet from 1952 to 1985, but most of this subsidence passed unnoticed by the casual observer. Likewise, the land over the central well field in Tucson has dropped about one foot from 1950, and in the 1980s was sinking about 0.8 inches per year—a subsidence rate that most Tucsonans would have difficulty pinpointing. On the other hand, if rates of subsidence increase with continued overdrafts, central Tucson could drop twelve feet by 2025, a scenario that would likely produce "differential subsidence," damaging homes, roads, and commercial property.[13]

In the middle basin in the 1930s, agriculture was under siege, at first assaulted by depression realities, then pressured by urban expansion. It remained a significant factor in the basin's draw on water, however, especially on the northern fringes of Tucson. During the bust of the depression years in 1931–1933, just north of the city in the Cortaro-Cañada del Oro district, the water table actually rose sixteen feet as production declined. Conversely, the war years and postwar boom had caused pumping to resume at full force and the decline of the water table commenced at an even greater rate. Approximately 750,000 acre-feet were removed from the aquifer through the war years and beyond. As a result, the water table had dropped an average of sixty-five feet from 1940 to 1956. The age-old pattern of ebb and flow, boom and bust remained in full view, although the modern realities created much more extreme variations in the pattern. The booms were bigger, and the demands placed on the river and its aquifer were greater.[14]

The huge overdrafts of groundwater affected the aquifer by lowering the water table and changing the nature of the underground flow. By the mid-1950s, hydrologists began noticing localized shifts in the direction and speed of groundwater migration. In the area of the Cañada del Oro alluvial fan, the water table was not declining from direct pumping so much as from the heavy pumping near the Santa Cruz River. Groundwater from the fan was flowing into the void left by the pumping near the river. In typical fashion, a "cone of depression" had occurred in the water table as water drained into the local void in the area around a well. Schwalen and Shaw noted that tracking of the groundwater from the Cañada del Oro fan indicated a new type of subsurface flow, a shift in ancient patterns of ground-water migration as wells and pumping affected the speed and direction of underground flow. As the cone beneath the Santa Cruz River floodplain increased in girth and steepness, it began pulling water in from surrounding areas, at times causing water to flow in odd directions. Above ground the river channel followed its traditional downhill slope to the north and west. Below ground, in places, water now oozed to the south and east, "uphill" in relation to the surface slope.[15]

New Deal programs during the depression mainly benefitted urban areas. This was true in the upper basin of the Santa Cruz River, where federal flood control projects had solved the problem of regular flooding in No-gales. In the middle basin, Public Works Administration (PWA) and Works Progress Administration (WPA) projects also concerned flood control along the floodplain as well as expansion of Tucson's water utility. University of Arizona hydrologist H. C. Schwalen recommended extensive efforts at bank stabilization, and New Deal agencies mostly followed his plans. Their efforts were largely successful in reducing any further widening of the channel, and the process of dumping old automobile frames into the river channel at the point of meanders contributed to the straightening-out of the river channel. But erosion continued nonetheless, and if the channel did not widen, it merely deepened, leaving the floodplain land in the former center of agriculture in the middle basin high and dry. As this was taking place, urbanization increased and the former agricultural land be-came valuable for residential and industrial uses. Much of Tucson's inner city rests atop the former cultivated fields of the old presidio. The arroyo facilitated drainage from the now-paved acres, and the generally reduced flows in the river through the 1920s and 1930s decreased concerns about catastrophic floods and property damage.[16]

The 1930s also brought federal assistance for expansion of the municipal water system. A $226,000 grant from the PWA financed 45 percent of the cost of the system's development. Included in the project were two 1-million-gallon "elevated storage tanks," a 7.5-million-gallon storage res-ervoir at 18th Street and Osborne Avenue, additional pumps and booster stations, and new water mains for the expanding areas of the city. As the newspaper explained, "Until the inception of the PWA program, expansion of the water system necessarily was along extremely conservative lines, although the additions included in the recently completed work had been contemplated for some time as the growth of the city made it more and more apparent that a considerable expansion soon would be imperative . . . [the project] left the system ready to meet the city's growth for the next ten years without requiring any additional major expenditures other than extension of mains to new territories."[17]

Although many New Deal programs benefitted the middle basin, a particular proposal fell flat in 1936. The depression era had witnessed the revival of the idea for a Sabino Canyon dam as a partial remedy for the economic downturn. Talk started in 1933, and in 1936 the Corps of Engineers formulated a plan for the dam with the proviso that the city and county share in the cost of the project. The Chamber of Commerce and city leaders backed the plan until the Corps came up with its estimate of the local contribution: $500,000. Since the chamber's fund-raising effort only came up with $600 for the project, the Corps dropped its plans for the dam.[18]

New Deal agencies also contributed to the development of recreation and tourism in the valley. Clearly agricultural uses for water dominated most of the valley, but the growth of urban areas and the increased interest in recreation and tourism in both the upper and middle basins indicated the future direction of the river's history. In addition to the traditional domestic, agricultural, and industrial uses—the three "C's"—surface water and groundwater now served recreation and tourism.

Whereas the Corps of Engineers' plan for a dam in Sabino Canyon failed to materialize, other New Deal programs helped develop the canyon as a favorite picnic destination for Tucsonans and tourists. The Emergency Relief division of the WPA built nine bridges over the stream in the canyon to allow easier access. As any visitor to the canyon can attest, the bridges are faced with attractive native stone, which was added by workers from the Civilian Conservation Corps.

The recognition that tourism could provide an economic boon to the valley had been apparent since the 1920s. In this, the valley shared in the rise of industrial tourism throughout the West, such as described by Hal Rothman in *Devil's Bargain*. In the upper basin, the Rancho Grande Hotel in Nogales, Arizona, advertized itself as a luxurious destination for wealthy sojourners to Mexico. One noted guest was film star Dolores del Rio. Dude ranches proliferated in the Santa Cruz Valley, seeking to tap the tourist market seeking realism, or at least verisimilitude, in their vacation retreats. In Tucson, the Sunshine Climate Club formed in 1922, advertising warm winter temperatures, lifestyles close to nature, and desert vistas with nary an obscured or mono-hued sunset. They also proclaimed the joys of shirt-sleeved golf in January. As the number of golf courses increased over the decades, green and lush amidst the desert terrain, they became a focal point for Tucson's water politics. All courses initially used groundwater to irrigate greens and fairways, and as the water table declined, critics began to question such recreational uses for groundwater. But in the 1930s, questioning such as this was still far in the future. The city opened the first of many public golf courses in 1936, and the fourth "C," climate, became firmly ensconced as a key selling point for the city.[19]

The Tucson Water Company maintains, somewhat arbitrarily, that 1940 marks the loss of balance between natural recharge of the aquifer and the draft of ground water by wells and pumping, as the severe overdraft of water resources in the basin begins in the 1940s. This assertion has made its way into the general understanding of the river in the basin, repeated in studies from the USGS and University of Arizona. For example, a recent university study published in 1999 stated: "Groundwater pumping near the Santa Cruz River began to affect the river flow, but not until the 1940s did this pumping finally cause the water table in the area to drop so low that the river flowed only during floods."[20] In other words, perennial flow lasted in the basin, at least in spots, well into the fourth decade of the twentieth century.

The assertion of lost balance in 1940 stems in part from the observation of changes in the riparian area south of Martinez Hill. As Betancourt explained, "In 1940, the water table near Martinez Hill and Sentinel Peak was still within a few meters from the surface of the floodplain. . . . The cottonwood galleries and mesquite bosques south of Martinez Hill, a popular picnic spot for Tucsonans in the 1930s and 1940s, died out leaving the floodplain treeless."[21] One particular description of the bosque in 1935 noted its apparent longevity: "In 1935 many a grand old patriarch still ruled here that had evidently already looked down on several centuries of desert droughts and savage storms. . . . Here there are trees of historic dimensions; the bole of one stately specimen . . . reached a girth of 13 feet six inches; and a diameter of more than 43 feet; while the height of another capitol-domed giant was calculated to be 72 feet."[22] It should be noted, however, that cottonwood trees and mesquite bosques had also recently occupied the land between Tucson and the San Xavier mission, but had been cut down and dried out since the 1880s. The land south of Martinez Hill was the last to suffer drops in the water table, not the first. Truly, the balance had been lost and the process of decline had been ongoing for decades prior to 1940.

Another aspect of the assertion of lost balance in 1940 stems from the recognition of how much water was pumped out of the aquifer after 1940. Through the 1930s and into the 1940s, the overall rate of increase in water pumping had declined due to depression pressures and declining rates of population increase. The water company had only brought ten new wells into operation during the 1930s and 1940s.[23] On the other hand, from 1940 to 1965, during World War II and the postwar boom, about 3.3 million acre-feet of water were pumped out of the aquifer under the basin. Over the same twenty-five years, as Tucson's population skyrocketed over 600 percent, from 30,000 to over 200,000, pumping for domestic uses rose from less than 7,000 acre-feet in 1940 to more than 54,000 acre-feet in 1965. And, of course, domestic use was small compared to pumping for agriculture. Irrigation volumes rose from 42,000 acre-feet in 1940 to 141,000 acre-feet in 1965.[24] This huge overdraft after 1940 makes any earlier overdraft pale in comparison. In addition, the overdraft of groundwater contributed to the spread of the arroyo upstream as far as Continental, and the deepening of the channel in the Tucson Basin. But the arroyo had been growing, and down-cutting, since the 1880s; it was not a process that began in 1940.

The rate of down-cutting was also influenced by the increasing use of the river channel as a landfill in the mid-twentieth century. The arroyo near Sentinel Peak became a notorious wildcat dump site. Did people know or care that they were depositing their trash near the site of the old San Agustín mission, or Warner's Lake? Other landfills were located at the former confluence of the river's west branch and mainstem, Congress Street, the former site of Silver Lake, the confluences of Rillito and Cañada del Oro with the Santa Cruz, and Marana. The construction of additional bridges across the arroyo also contributed to the down-cutting, as did sand and gravel mining, and the dumping of overburden during road construction.[25]

The modern hydraulic realities of life in the middle basin did not occur out of thin dry air in 1940. The modern vision had its origins far in the past and only became acutely visible in the first half of the twentieth century. Much akin to the vast underground aquifer, invisible to so many for so long, the potential hazards of relying on groundwater for the valley's domestic and commercial water supply remained hidden and absent from the thoughts and concerns of most valley inhabitants. Even when an awareness emerged, such as G.E.P. Smith's perception of the problem with overdrawing groundwater supplies in 1910, the awareness was tempered by the modern, progressive optimism that promised new sources and advanced techniques for harvesting water supplies.

The modern perception was fully ensconced in the valley by 1940, but this did not mean the archaic, or premodern, view of the river had disappeared completely, vanquished by the new powerful ideology. Paradigm shifts do not turn on and off like a light switch. Premodern perceptions intrude even into the engineers' reports of "live" water. Why not "useful" water, or "workable" water instead? Clearly the engineers meant that surface water was less troublesome than groundwater: it was more readily usable, less expensive to acquire, and easier to manipulate. But the term used to designate this usefulness carried with it clear anthropomorphic connotations. It ascribed a poetic truthfulness to clear flowing water. More than simply useful, it was alive and life-giving.

But the future course was set by 1940, and it was not a future to be borne on live surface water. The "vast untapped resources" of groundwater would assure the valley's prosperity. Grandiose schemes to import surface water across the desert from distant watersheds remained pipe dreams in 1940. The Santa Cruz Valley was on its own. Now it was the aquifer, not the surface flow of the river, being asked to support the future generations in Tucson for centuries to come.

Business leaders of Tucson looked forward to the decade of the 1940s as a prosperous and expansive period. An extensive supplement to the *Arizona Daily Star* during the annual February rodeo hoopla, boosted the city, valley, and state with unabashed enthusiasm. Later in the year, the city's tourist promotion organization, the Sunshine Climate Club, announced with glee the numbers of visitors (1,404) and new families (240) that had arrived in Tucson during the first eight months of 1940. The club leaders predicted that the coming winter would witness the biggest snowbird season ever.[26] Most of the 1,014 visitors to Tucson counted by the SCC in 1940 had been winter visitors and spring sojourners—few tourists were willing to brave the blazing summer temperatures—and high hopes reigned that the winter of 1941 would bring an even greater snowbird migration.

Of course the Japanese attack on Pearl Harbor would interrupt the club's tourist campaigns, but the war did bring tens of thousands of new visitors and residents attached to military units assigned to area bases. So even though the tourist predictions fell by the wayside, the war years and post-war boom ushered in a new era of prosperity in the middle basin. The river and aquifer's circumstances changed markedly as a result.

The river had always been the center of life in the valley, but given the river's fitful nature of modest surface flows and periodic shortfalls, human society in the valley had frequently acquired a paranoid tinge. If summer rains were a week or two late in coming, or if the first winter storms bypassed the valley, worried pronouncements of water famine and drought would circulate among the denizens. So paranoia about water was nothing new to the middle basin when it surfaced again in December 1941; it simply appeared as a war-inspired variation on an old theme.

After Pearl Harbor, as civil defense officials and city law enforcement discussed the advisability and management of a citywide blackout, the mayor ordered guards be stationed and a fence built surrounding the water storage tanks and reservoirs in the city. The supposition, correct in its basic elements, was that Tucson was extremely vulnerable to any assault on its water resources. Should the water system break down, or suffer irreparable harm through sabotage, life in the desert valley would become virtually impossible for its inhabitants. But did the mayor and other officials actually think that Japanese or Nazi sympathizers would attack Tucson's water system? Davis-Monthan Air Field east of town—certainly a potential target for saboteurs—had its own wells and pumps, and so an attack on Tucson's water system would have no direct effect on the military's presence in the valley. But such was the fear in those days after the Japanese attack on Pearl Harbor, and thus the river and its life-giving water played a role, or at least a bit part, in Tucson's version of war hysteria.[27]

As to the city's water system and the middle basin's utilization of the river and aquifer, the war years saw relatively few new developments beyond increased demand due to the growth of the city. The federally subsidized expansion of the water system in the closing days of the New Deal had provided enough added capacity to see the city through almost a decade of growth. During the war, much of the city's development related to the increasing military presence in the valley, as Davis-Monthan Air Field, Ryan Field west of town, and Marana Air Field north of the city, trained thousands of Army Air Corps flyers. Also, Consolidated Aircraft Company operated a plant at the city airport, modifying B-24 bombers originally constructed on the West Coast. These military operations brought many new residents to the middle basin, and so contributed to the increased draw on the aquifer. The operations also introduced a new factor into the river's history as industrial pollutants related to the aircraft industry, carelessly disposed of, started their inexorable trickle into the water supply. It would take decades for the solvents and oils to show up in city well water, but the process had its start in the years of World War II.[28]

As the war years gave way to the postwar boom in Tucson, the harsh realities of the river's limits came into high relief. In order for the Tucson Basin to continue its remarkable growth, and to expand the agricultural production in both the upper and lower basins, more water had to be found. Without deeper wells or new sources the boom times would come to a disappointing halt. But as the pumps whirred and the water table plummeted, undeniable concern arose in the city's legislative delegation and among other legislators representing agricultural districts. In 1945 the legislature passed a law requiring the registration of large wells (100 gallons per minute or larger), and in 1948 passed another law that allowed for the creation of "critical ground-water areas." If the state land commissioner declared an area "critical," no new land could be brought into production through irrigation (existing production and pumping would continue unabated). As of 1952, four agricultural areas had been declared critical, two in the Salt River valley and two in the Santa Cruz River valley. The critical areas in the Santa Cruz Valley were downstream from Tucson in the lower basin.[29]

Manifestations of the concern for water supplies also appeared in the proliferation of studies and scientific reports on the river valley's water supply. Mostly conducted by University of Arizona hydrologists and USGS engineers and scientists, these reports offered both dire warnings and optimistic prognostications. Continuing in the tradition of G.E.P. Smith, the reports provided both scientific data and boosterish pronouncements. Of benefit to later observers, the reports also provide an accurate account of the water table's steady decline through the middle decades of the twentieth century.

The first of the major postwar scientific reports, a USGS study of the water supplies in southern and central Arizona, appeared in 1952. The report was supervised by Leonard C. Halpenny, chief district engineer of the USGS, and included analyses of water resources along the entire length of the river. Halpenny and his staff of hydrologists reported huge overdrafts of the aquifer in all the basins. By far the greatest overdrafts were in the lower basin, as has been previously noted.[30]

In the upper and middle basins, the primary source of water was recharge, since surface flows in the early 1950s were all but nonexistent in any meaningful or useful form. P. W. Johnson, the hydrologist who compiled the upper and middle basin data, figured that annual recharge of the aquifer was in the neighborhood of 125,000 acre-feet along mountain fronts, main stream channels, and from irrigation runoff. More water traveled through the basins as "underflow," but the total supply remained far in arrears. From 1947 to 1951 an average of 30,000 acres of irrigated agriculture used 105,000 acre-feet of water. In 1952 agricultural production had increased to 35,000 acres, and relied on 1,000 wells drawing water from an average 150 feet. Nogales and Tucson were also drawing water out of the aquifer, but these municipal uses were much less than the total for agriculture and industry. The total draw from the aquifer in 1951 had been 191,000 acre-feet, with 37,000 acre-feet going to the two cities (Tucson, 28,000; Nogales, 9,000). The overdraft, or "net annual loss from storage," averaged 55,000 acre-feet per year from 1947 to 1951. Johnson reported the overdraft without hyperbole or fanfare, remarking only that "the water table has declined steadily in most of the basin since 1939." Of course in some portions of the basin, namely where the first wells in the middle basin had gone into production in the 1890s, the water table had been declining for much longer. Nonetheless, Johnson's remark indicates that the phenomenon of overdraft had spread beyond the Tucson Basin to the south.[31]

Another report, much more passionate than Halpenny's USGS report, came out the following year in 1953. Boosters in the Chamber of Commerce attempted to keep the postwar boom times rolling with "The Truth About Water in Tucson," a pamphlet composed of reports by Phil Martin, the water system manager; a Phoenix-based engineering firm's report; and a letter from Leonard Halpenny. The two reports and letter applied a bit of spin doctoring to the descriptions of groundwater pumping in the Tucson Basin. At the rates recorded in 1951, an annual overdraft of 82,000 acre-feet had resulted in an average drop in the water table of three feet per year. At that rate, the reports said, Tucson could count on an eighty-year supply, or fifty-year supply "at reasonable pumping lifts." The fifty-year projection to 2001 must have seemed safe enough in 1953, perhaps less so now. The reports also characterized nearby water supplies that Tucson could aspire to develop: the upper Altar Valley, twenty-six miles west; the proposed dam on the San Pedro River at Charleston; agricultural lands in the upper Santa Cruz Valley; and CAP water, whenever it became available.[32] Despite the best efforts of the USGS and University of Arizona surveyors, however, no silver bullet solution for the intractable circumstance of aridity in the basin ever appeared.

The 1952 study supervised by Halpenny had reported overdraft of the aquifer without much editorial comment. Such was not the case in another report by University of Arizona hydrologists published in 1957. Schwalen and Shaw's study began with an explicit statement of the problems of overdraft, and the "imperative" need for information on the basin's water resources: "The increasing demand for water in the Tucson area made a study of the effects of pumping upon the ground-water table imperative. Officials of the City of Tucson and Pima County realized that accurate and detailed information of changes in ground-water levels must be available."[33] The report was bound in book form, with a cover by noted Tucson artist Ted DeGrazia, and marked a move toward dissemination of such hydrological information to the general public. Hard choices loomed in the future, and the competition for water was destined to become ever more difficult: "If the city continues to grow at its present rate, some water will have to be diverted from agricultural use to meet future municipal needs, or new sources must be found." The scientists went on to warn Tucsonans that development of mining interests south of the city would be likely to increase the demand on water, and thus the old mantra of new sources became doubly important: "The possibility of capturing flood water for recharge is being investigated by this Department and other agencies. Importation of water from the adjacent San Pedro or Avra Valleys is also being considered as a source for future water supplies." In part the report served the interests of the politicians advocating pursuit of new sources, but the report also served notice that shifting priorities of water use were not far in the future. Missing from the 1957 report was any call for conservation along the lines of the later "Beat the Peak" program (to be discussed in chapter 11). Since agriculture still used far more water than the urban areas in the valley, the main culprit in the narrative of declining water tables was irrigated farming.[34]

The 1957 report confined itself to the upper and middle basins. Most of Schwalen and Shaw's study described the state of the water table in relation to the level of agriculture in the region. Overall, the report claimed a limited decline in the water table for the Tucson Basin prior to the 1940s, and as such stands as one of the main pillars of the "lost balance in 1940" argument. But within the assertion of balance is the reference to more site-specific declines in the aquifer in the 1930s, which basically show that the water table declined wherever and whenever groundwater pumping began. In the Santa Cruz "bottomland" between the mission and the city, "negligible" declines of ten or twenty feet (negligible only in relation to the much larger declines of later decades) was sufficient to eliminate any vestige of surface flow along that reach except during times of flood. Wet years, such as 1954–1955, produced some recharge of the water table, but overall the trend was decline. The report provided a contour map showing the decline of the water table throughout the upper valley, and provided well log data for 286 wells in the upper and middle basins. Over 300,000 acre-feet of water had been removed from the "ground-water reservoir" since 1947.[35] To place this draft of water in perspective, lower basin farmers had pumped 344,000 acre-feet of water in the single year of 1941, and by 1957 were pumping well over one million acre-feet every year. This is not to say that the overdraft in the upper and middle basin was less than serious; rather, the one level of overdraft was huge, and the other was simply beyond comprehension.

In part the surveys and reports had always been meant to support the acquisition of additional water sources, at first within the basin and eventually without. The search for solutions for water scarcity in the semiarid region never lagged, and at least rhetorically, the search for additional water sources outside the Tucson Basin came to fruition in the late 1950s. In November 1959, the city manager, Porter Homer, described the city's plans to develop "hitherto untapped water sources." Homer explained that the city was spending $10 million to sink new wells in the Santa Cruz Valley to the south of the city, and in Avra Valley to the west and north. These wells were the result of successful bond elections in 1952 and 1958, which financed the development of new well fields. The "Santa Cruz Well field" south of Valencia Road added thirty-four new wells to the city's system. These were the wells that depleted the water table to such an extent that the remaining mesquite bosque south of Martinez Hill dried out and died off, noticed by Tucsonans in the early 1960s.[36] Homer also discussed plans for the CAP and Tucson's efforts to claim water from the proposed dam on the San Pedro River, even though the water from the Charleston dam would not be needed for "30 or 40 years."[37]

While seeking to control additional sources beyond the Tucson Basin, city leaders jealously guarded the water resources within the basin, and kept vigil against any perceived siphoning. By 1954 the city's interests no longer included agriculture within the basin, and so agricultural usages for water now posed a threat. In this light the city opposed—generally without success—private wells in the valley, especially if those wells serviced agricultural pursuits.[38]

As the city's opposition to agricultural wells indicates, the turn in water politics after World War II was pronounced. Controversy and scandal had occasionally centered on the water department in Tucson, but basic consensus held firm that agriculture was central to the city's prosperity, and never was a proposal for a new agricultural development anywhere in the region opposed on grounds other than financial feasibility. Of course there was also consensus that the limited water supply made domestic urban uses problematic, especially during droughts or system shortfalls. At those times, as water customers in Tucson suffered through outages or low pressure, water politics emerged in sometimes rancorous fashion.

But the passion surrounding water politics seemed to ratchet up during the postwar boom, and such seemed to be the case in 1949, when complaints of low water pressure in several neighborhoods generated a $2.5-million bond election to finance construction of larger water mains. The bond issue seemed headed for quick and easy approval. The expansion of the system during the New Deal era had served pretty much as promised, without major additions for about ten years. Thus city leaders had not felt the voters to be overburdened with tax increases to service water bonds. Then, two days before the election a group calling itself "the tax committee of the chamber of commerce" mounted a last-minute opposition campaign. The group took out newspaper and radio ads complaining that the half of the $2.5 million slated for general obligation bonds (the other half were revenue bonds) would unnecessarily raise residents' tax burden. City leaders thought the last-minute campaign confused voters, since the water department and city leaders had no opportunity to counter the criticism of the proposed general obligation bonds. The inability to mount a public relations campaign countering the claims, plus the low turnout of only 10 percent of eligible voters, caused the general obligation bond portion of the election to lose by 347 votes (940 to 593).[39]

Democrats had maintained control of the city council in April 1949 despite a similar tactic by Republicans. The night before the election, Republicans announced a full slate of write-in candidates. Their hope was that Democratic turnout would be low, since the Democratic candidates had assumed to be running unopposed. Given a low-enough turnout, the Republican write-in candidates might stand a chance of winning. Only a frantic effort by Democrats to get out the vote on election day averted a Republican victory. Despite this forewarning, the city council had not expected the opponents of the water bonds to use similar tactics in the special election.[40]

Most water bond issues passed without such electoral difficulties. Voters approved significant expansions of the water system in 1952 and 1958 that added wells south of the city along the Old Nogales Highway and in Avra Valley. But just as bond elections could face opposition, water rate increases also were subject to volatile water politics. A rate increase in 1952 passed with little fanfare and no opposition. In 1952 the 13 percent increase in rates accompanied a $5.5 million water bond election, and neither generated noticeable resistance. One reason the rate increase sailed through was simply that it was the first increase since 1925 (incorrectly stated by the city as the first increase in 29 years—actually it was 27 years). However, the city's effort to raise water rates again in 1959 generated angry opposition. The city proposed raising rates by about one third so as to provide "for expansion of facilities to serve more than 600,000 population by 1995." The passage of the rate increase on August 11, 1959, came after a heated debate between the Democratic mayor and Republican councilmen (the city council in 1959 was divided, four Democrats to three Republicans). The debate presaged opposition to the rate increase from the Property Owners Protective Association, a group of disgruntled residents seeking to call a referendum election to turn back the water rates to the 1952 level. The city resisted the effort to call a referendum, and the controversy dragged on until a court injunction in April 1960 ended the water rate election.[41] This opposition to a water rate hike foreshadowed another such electoral battle almost twenty years later, as a proposed revision of water rates brought the "slow growth" movement in Tucson to a startling halt—fodder for the next chapter.

In another manifestation of water politics, annexations of subdivisions outside the city limits in the postwar boom period caused the water department to purchase many private water companies. This process was generally expensive and often contentious, and also resulted in repeated doublings of the water department's size. The water company announced with pride their growth spurts, doubling in size from 1945 to 1955, and doubling again from 1955 to 1957. In the two years from 1955, the city had acquired five private water companies. In an earlier example of such an acquisition, in 1949, a prominent and politically connected developer, Evo DeConcini (Democratic Arizona Supreme Court Justice), asked the city to pay $15,000 for a private water system he had installed in one of his subdivided developments the city had recently annexed. The water department maintained that the system was worthless, but the city nonetheless offered De-

Concini $2,500. The judge sought arbitration to arrive at a "fair" price between $2,500 and $15,000, but the city council refused.[42]

The progress toward a postmodern perception of the river was well underway by the 1950s. In much of the valley the river had long since ceased to flow and only in the form of dark, muddy, swirling floods did anything resembling a river move down the channel. Mention of the quality of the floodwater is necessary to indicate how far removed the river was from most valley residents' perception of water, especially in Tucson. To most residents, water was clear, cheap, and readily available at the turn of a tap. The closest thing to that water in their day-to-day consciousness might be the occasional rain shower, but certainly not anything traveling down the river channel. Some might think that floodwater was wasted, coursing swiftly down the channel to dissipate in the Santa Cruz Flats, or possibly flow into the Gila and be lost forever. The CAP was still a panacea, distant and alluring, and seemingly a political impossibility. The river channel had become a trash dump and an inconvenience to urban development, and pressures had mounted to build on the floodplain.

As human engineering of the river and floodplain increased through the middle decades of the twentieth century, it seemed that only the occasional flood needed to be controlled before more houses could be built in close proximity to the arroyo. The expectation of occasional floods seems to have become accepted as the "nature" of the desert river. In 1960 few inhabitants of the river valley would have believed that as recently as eighty years earlier the river ran perennially in many stretches along its 210-mile course. Two cities had come to inhabit formerly lush areas on the river and its tributaries. Tucson had gobbled up much of the floodplain and Nogales had grown up directly on top of the tributary canyon's floodplain and shallow aquifer. The Nogales pump house at the mouth of Proto Canyon finally ended the perennial flow along that stretch of river in the 1960s.

Generally low-flow conditions from the 1930s through the 1960s, especially the severe drought in the 1950s, allowed urbanization in the Tucson Basin to proceed in proximity to the floodplain with a casual attitude toward potential floods. The peak flow during the winter of 1914–1915 remained a distant memory, and no such flood rolled down the river again until the summer of 1961. Other floods of similar magnitude passed down the channel in the 1960s, but these flows occurred within a context of "normal" floods that seemed to conform to residents' notions of the usually dry river channel only occasionally and spectacularly flowing through Tucson. For the most part the floods had been safely contained within the incised arroyo.[43]

To increasing numbers of residents of the valley, the water in their glasses at dinnertime bore little relation to the river that had given birth to their community.

PART III

Postmodern

11

Preservation and Conservation
1960–1983

FLOODS ARE A FEATURE OF LIFE IN the river valley, and the floods of 1977 and 1983, and to a lesser extent the flood of 1967—the greatest floods in the river's history—brought that simple fact into prominent relief for valley residents. Nogales suffered major damage in the 1977 flood, and the 1983 flood set new records for discharge and damage. Once again the river flowed over its entire course, reminding residents (and historians) that a narrative of the river's history restricted to one or two basins would be incomplete. The floods also reminded residents that scientific notions of river management and flood control could wash out as easily as a bridge abutment.

Although the three basins continued to develop along their own trajectories, the floods tended to blur distinctions between them. Regional and national developments also affected the entire valley, becoming clearer and more direct influences on the river in the last thirty years of the twentieth century. Of course, the Great Depression, New Deal, and World War II had affected the valley, but the presidential initiatives and federal programs in the 1970s and 1980s increased the level of governmental intrusion. It is perhaps ironic that as forces beyond the valley began to exert their greatest influence, the river had already ceased to exist through much of its former course, except during spectacular floods or as an invisible, underground resource.

The railroad had dragged the little river into a much larger world at the dawn of the twentieth century; as the twentieth century rushed to a close, the little river no longer flowed much at all. One even hesitates to call it a river, related in fundamental quality to the historic stream. Coined "ever a dwindling stream" in 1910, by the end of the century the Santa Cruz River, especially in the middle basin, was comatose if not already dead.

Whether the river was deceased or merely somnolent, the politics surrounding the river and its aquifer continued to bark and grumble. Water politics reached towering heights of rancor and the deepest chasms of divisiveness in the last decades of the twentieth century. And once again, Tucson comes to dominate the narrative. The new realities of water politics in the middle basin, namely huge increases in groundwater pumping with the panacea of the CAP shimmering like a mirage in the distance, caused political and bureaucratic careers to rise and soar, and to crash and burn.

One reason for the discord was that from at least 1977 up to the present, any discussion about water resources in the valley likely would bring residents of Tucson, Nogales, or Tubac—ranchers, Native Americans, miners, golfers, or suburbanites—into a hopelessly convoluted debate because their views of the river and the valley's water resources arise from at least three distinct perceptions: archaic, presuming water to be living, imbued with spiritual or mystical power, and worthy of reverence; modern, seeking the most efficient use of the resource and placing faith in science and technology to solve the region's endemic scarcity of water; and, postmodern, completely disassociated conceptions of the resource from the sparkling convenience trickling from chrome taps and brass valves. The resolution of water politics in the valley consistently runs afoul of these conflicting perceptions, making consensus building a daunting task to say the least.

In the upper basin, the 1960s proved a pivotal decade for the river as the urban centers of the twin Nogaleses finally created a persistent and inexorable decline of the water table. The wells and pumps in the service of expanding urban populations at last ended the perennial flow in those stretches of the river upstream from the two cities and at the mouth of Proto Canyon. Then, in an ironic reversal, surface flow returned to a stretch of the river downstream from the international sewage treatment plant, just as other stretches of the river turned dusty from declining water tables.

Riparian areas and surface flow in the upper basin were under severe assault in the 1960s. Only in the San Rafael Valley and along Sonoita Creek near Patagonia did perennial surface flows persist. In the high valley ranching continued in long-established patterns and the stream benefitted from the generally enlightened grazing practices. Only dirt roads serviced the area, and so development pressures felt elsewhere in the river valley largely bypassed the grasslands. The riparian area south of Patagonia—a favorite picnic spot known as Blue Heaven by locals—remained into the 1960s, and was eventually preserved from development by The Nature Conservancy in partnership with the Tucson Audubon Society. In February 1966, the environmental groups purchased the first 312 acres and established the Patagonia-Sonoita Creek Preserve. In 1970 the area became a National Historic Landmark, and since 1970, the preserve has grown to 750 acres through "conservation easements and donations." Now the stretch of Sonoita Creek within the preserve is a renowned bird sanctuary and wildlife refuge, noted for the 290 bird species and four native fish species that inhabit it.[1]

This example of a riparian area preserved by environmental groups serves as a good indication of the continuing relevance of the archaic perception of the river. By this I certainly do not mean to suggest that the environmental activists in the Nature Conservancy or Audubon Society necessarily shared the worldview of prehistoric Hohokam, colonial Spanish, or present-day Native American inhabitants of the valley. As I define the archaic perception, it merely requires that a respect and reverence be directed toward the flowing stream. Many cultural variations may occur within this broad definition, which is useful for understanding the current water politics in the valley. Preservationists clearly value the "living" stream for the life-giving force it brings to the surrounding ecosystem. Once preserved, the goal is to perpetuate the natural setting, most likely with some concession made for observation and personal experience: birdwatching, backpacking, picnicking, or some other sort of Muirish activity. This valuation of the stream, I presume, differs from the kind of reverence directed toward the river by, say, the Hohokam.

Preservationists in the 1970s often acted with a sense of urgency created by the startling decline in surface flows. As they pointed out, when the streams "died," much beyond the flow of water was lost. Native fish species from minnows to trout disappeared. Countless other species of swimming, crawling, and burrowing critters—tadpoles and frogs, crayfish, salamanders—all gone. Cottonwood, willow, walnut, oak, and hackberry—dead. With the trees gone, hundreds of bird varieties, from the smallest hummingbirds and songbirds to the largest and most startling raptors—vanished. Mammal species—pack rats and mice, cottontail and jackrabbit, skunk, raccoon, coatimundi, bobcat, grey fox, coyote, varieties of deer— were much diminished or absent, except for those animals most adaptable to human society. The list goes on, thousands of species of plant and animal life inhabiting the riparian ecosystem and dependent on the surface flow of water, on the verge of death if not extinction.[2]

Riparian zones closer to urban centers seemed doomed. Even agriculture, long the mainstay of economic development in the valley was under siege. In perhaps the clearest expression of the incompatibility of agricultural and urban uses for water, the Mexican government in 1965 passed a law prohibiting any further agricultural development on the Santa Cruz River in the vicinity of the Mascarenas family's holdings. Specifically, the water in the river and its aquifer was to be conserved for the growing cities of Nogales, Sonora, and Nogales, Arizona, and starting in 1965 the Mascarenas family ceased farming on the bottomland.[3]

The conservation policy in Mexico lasted about sixteen years, when water politics took a decidedly populist turn. On the Mascarenas Ranch the land-reform movement initiated by President Lopez Portillo resulted in 1981 with the seizure of almost all of the family's holdings. The land was then dispersed to 425 landless peasant families who commenced farming on the Santa Cruz River bottomland. The Mascarenas family protested this redistribution of their land and appealed the decision in the Mexican courts. In the meantime, the *campesinos* (farmers) sank more wells into the river's aquifer, despite federal law prohibiting such practices. In this case two federal initiatives conflicted and in the process, more water was withdrawn from the river's aquifer. One result of the confused federal policies was a lack of support for the new farmers on the river. Meager federal assistance coupled with the demands and uncertainties of agriculture in general caused the experiment in land redistribution to fail. As Manuel Mascarenas Manriquez reported in 1991, "The original 425 *campesino* families have dwindled in number to approximately 120 people. Most of the *campesinos* became disillusioned and abandoned the various ejidos [communal farms]; many crossed illegally into the United States in search of a better life."[4]

In addition to national policies in both countries, international politics also played a role in the economic development along the border. In the 1970s the U.S. and Mexican governments agreed to establish the *maquiladora* system of twin plants. U.S. manufacturers maintained parts warehouses and shipping centers on the Arizona side of the border, while operating assembly plants on the Mexican side, taking advantage of the cheap labor. Tens of thousands of aspiring workers from rural areas throughout Mexico flooded into Nogales, Sonora, hoping to land a job in one of the U.S. factories. Most were disappointed, but the hopeful kept coming, and the draw on the water resources increased, as the strain on the sewage system became acute.

The cities had been growing through the 1950s and 1960s, even before the twin-plant system attracted additional residents, and all through the decades of growth, the cities' water systems had expanded in an effort to maintain pace. The Nogales, Arizona, system in the mid-1960s added several wells in the river channel near the original infiltration galleries and pump house. The wells tapped the full extent of the shallow and narrow aquifer, drawing water to depths of 100 feet, but tests indicated that only the top thirty to fifty feet of the aquifer—loose gravel and rocks allowing groundwater to move freely—allowed for quick recharge. Below fifty feet the aquifer was composed of clay, and recharge rates significantly declined. In one test, the top thirty-eight feet of a well produced 850 gallons per minute, while the well below thirty-eight feet only produced 30 gallons per minute. In addition to new wells, the city began eyeing other water sources, acquiring surface water rights in Lake Patagonia, a recreational lake formed by a dam on Sonoita Creek constructed by the state in the early 1960s. The Nogales water managers also began drumming their fingers along with other municipalities, waiting for CAP water to arrive. Across the border, the Nogales, Sonora, system continued to struggle with the increasing population. Since many of the residents on the hillsides and up the canyons had no water hook-ups, trucks delivered water on a (hopefully) daily basis. The Nogales, Sonora, system also began eyeing water sources beyond the valley, particularly, water from the Magdalena watershed to the south of the Santa Cruz River valley.[5]

Downstream from Nogales, surface flow returned to the river after a dry hiatus of decades without much initial fanfare or commotion. This reversal of fortune for the river had as its cause two factors, according to Douglas Cumming. The first was increased runoff from higher precipitation amounts, and that floods, especially in 1967, had recharged the aquifer and spread thousands of cottonwood seedlings along the stretch of the river north of the Sonoita Creek confluence. The second factor was the consistent flow of treated effluent from the international sewage treatment plant. Cumming was quite correct in assigning much of the river's reversal to the discharge of treated effluent into the river channel. A similar scenario played out on a smaller scale in the middle basin, as effluent flows from sewage treatment plants in Tucson returned a perennial flow to the basin in 1966.[6]

Cumming also provided a vivid description of the flood in 1967, which he used as an example of the higher runoff amounts aiding in the river's return:

> In December 1967 I had a chance to view one of the two biggest Santa Cruz River floods in memory. I sat on my horse on a hillside observing the river. The Santa Cruz was more than a quarter of a mile wide, all fast moving flood water. Towards the center of the flood, waves were cresting three or four feet high, and large uprooted trees were rolling and bobbing along. Across the river a fifty-yard stretch of railroad had been washed out and a torrent of water was gushing through the cut.[7]

Although the higher precipitation amounts that had generated the flood were clearly a factor in the recharged aquifer, the effect on recharge of the flood itself was less significant. Most of the floodwater passed quickly through the channel, with little chance to percolate into the aquifer. On the other hand, floods like the 1967 event occurred when the ground was saturated from days or weeks of steady rain, with the flood occurring after a final deluge slid off the saturated surface. As such, a flood can be taken as an indication of significant recharge, if not from the flood itself, then from the preceding rainfall that contributed to the flood.

In the lower basin, farming continued apace into the 1970s, depleting groundwater at a remarkable rate. In 1950 there were 408 farms in Pinal County cultivating 210,011 acres that relied on groundwater only, pumping steadily from the Santa Cruz and Gila River aquifers. The farms in the lower basin were using 1,550 wells with an average pumping lift of 174 feet. However, this does not account for all the agricultural land in the county, since many farmers had access to the distribution systems operated by irrigation districts and reclamation projects. Considering all sources of irrigation, there were 867 farms in the county cultivating 358,383 acres through irrigation.[8] Thus, over 58 percent of the irrigated farmland in the county was relying solely on groundwater in the early 1950s. The region had been designated a critical groundwater area in 1948, and the number of acres irrigated solely by groundwater declined gradually over the ensuing decades, even though the total number of acres irrigated by all sources increased. The following table shows the peak of agriculture in the 1950s and decline in the number of farms and acres relying on groundwater only into the 1980s.

Pinal County

	1950	1959	1982
Farms relying on groundwater only	408	320	235
Acres irrigated by groundwater only	210,001	154,184	112,418[9]

However, the total number of irrigated acres in the county increased into the 1960s, as the number of farms generally decreased.

	1949	1954	1959	1964	1974	1978	1982
Farms relying on irrigation	1,058	867	625	538	409	406	454
Acres irrigated by all sources (in thousands)	266	358	1,730	1,650	912	761	1,110[10]

Agriculture remained the dominant economic force in the lower basin in Pinal County. Cotton, grain, and eventually pecan groves, continued to rely on the aquifer and other precious, dwindling water sources. The CAP was awaited with keen anticipation in the lower basin, but it remained a triple-sided question for area farmers. First, of course, was the basic question whether it would ever arrive. Second, if and when CAP water did arrive, would it cost so much that individual farmers and irrigation districts would go broke purchasing it? The third question regarded the water quality. At issue was not so much that crops would wither under it, but some questioned whether the land would become quickly and irreversibly salty or alkaline. In the meantime little changed. Hot, dry, and sunny, aridity dominated the lower basin and groundwater levels continued to drop.

As Tucson grew through the 1960s, city leaders continued the multi-faceted program to protect and expand the city's water supply in the middle basin. The effort was pursued by political leaders of both parties and was generally, though not always, backed by broad political support. Through the 1960s and 1970s the city bought about 10,000 acres of farmland in Avra

Valley and shifted the output of twenty-seven wells previously devoted to agriculture to the urban domestic supply.[11] Also in the 1960s, the city began acquiring private water companies outside the city limits in an effort to control and regulate water usage throughout the basin. Difficulties in this effort arose both from an engineering and political perspective. The private systems often did not connect easily to the city's mains and maintaining consistent water pressure proved a challenge. Also, the county customers had no voice in city elections, and often felt ignored when city staffers discussed water policies or placed water issues before city voters.[12]

The acquisition of private water companies ran the gamut from difficult, expensive, and litigious, to peaceful, welcome, and efficient. Many residents welcomed the city services since private companies sometimes provided haphazard service. On the other hand, competently run private systems sometimes survived the annexation process. The Amphitheater and Flowing Wells areas entered the city in the early 1960s while maintaining their independent water systems.[13]

Development stretched to the farthest fringes of the basin by the 1960s. The stretch of valley from San Xavier to Canoa had been a riverless jornada for centuries, the vacant buffer between the upper and middle basins. But pump technology had brought cultivation to that area in decades past, and in the postwar boom, the mining industry joined the increasing development. Tucson was the nexus for industrial and agricultural activity, and as the stretch of river south of San Xavier developed, Tucson's domination of the middle basin began to spread beyond traditional boundaries. As the mass of Tucson's economic presence increased, the pull of its orbit and its domination of the valley extended into the other basins to the north and south. This trend will be explored in the next chapter, but its origins were clear by the 1960s. As the open-pit mines south of Tucson reached their peak production during the 1970s, water use increased to more than 50,000 acre-feet a year. By the early 1990s, as copper prices dropped and mines curtailed output or shut down altogether, water usage dropped to just over 30,000 acre-feet.[14]

Within sight of the tailings dumps at the mines, the retirement community of Green Valley sprouted in the 1960s adding another urban and golf-course demand to the valley's water resources. Within sight of Green Valley, cultivation along the river near Continental shifted from cotton to pecan groves. Between Continental and Sahuarita, about 7,000 acres of pecan trees required about 30,000 acre-feet of water a year.[15]

In all cases, be it mines, pecan groves, or golf courses, Tucson provided the cultural and political center of gravity. Water politics in the northern fringes of the upper basin had fallen into the orbit of Tucson.

Scientific and scholarly studies continued to provide basic information about the river, generally in support of the political efforts to expand the water supply and develop the valley. E. S. Davidson summarized ground-water withdrawals in the Tucson Basin from 1940 to 1965 in a USGS report published in 1973. One of the significant conclusions of the 1973 report was that the actual withdrawals from the aquifer were probably smaller than the calculated volumes indicated. The assumption had been that declines in the aquifer throughout the middle basin occurred as they did in adjacent wells, but many of the well sites were isolated, with wide distances at times separating data points. What Davidson suggested was the commonsensical notion that the decline in the aquifer between widely spaced wells was probably less than in adjacent wells. Nonetheless, the withdrawal from the aquifer was immense. Pumping in 1965 totaled 177,000 acre-feet, and overall during the twenty-five years from 1940 to 1965, previous studies had calculated 3.3 million acre-feet removed from the aquifer. Here is where Davidson critiqued the previous calculations, and stated that his analysis showed the actual amount "did not exceed 2 million acre-feet." In another assertion of miscalculation, Davidson suggested that the amount of water used within the City of Tucson may have been overestimated in the years 1945–1955 and 1957–1958 because large areas had been annexed into the city. The most reliable data for municipal water use was for the year 1965, since "most of the pumpage was metered and a special census was taken." As a result, Davidson calculated 175 gallons per day per person for a population of about 235,000 in 1965, and that the overall amount of water pumped for domestic use had been 46,000 acre-feet.[16]

Another of Davidson's observations was that the municipal and industrial use of water was increasing at such a rate that they would, "within a few years," surpass the amount used for irrigated agriculture. Where pumping for the city had been less than 7,000 acre-feet in 1940, it was over 54,000 acre-feet in 1965 (or was it 46,000 acre-feet? Davidson cited "slightly more than 54,000 acre-feet in 1965" for "public supply" on page thirty-seven, and 46,000 acre-feet in 1965 on page forty). Industrial use of water had risen from 1,000 acre-feet to 18,000 acre-feet over the same period. Irrigation uses had reached a maximum of 141,000 acre-feet prior to 1958, but the level of irrigation declined from 1958 to 1965 as agriculture fled the city and expanding urban fringe.[17]

Finally, Davidson noted that the decline in the aquifer was not uniform throughout the Tucson Basin. As would be expected, the greatest declines were occurring in the areas of greatest pumping, but in some of those areas recharge of the aquifer occurred with greater efficiency, thus mitigating somewhat the effects of overdraft. Such was the circumstance in the area of Cortaro, "where the aquifer receives recharge from all the major tributaries and from the Tucson sewage system effluent." The areas of greatest water table decline were "along a 12-mile reach of the Santa Cruz River near Sahuarita and from the eastern part of the urban area to a few miles southeast of the confluence of Rillito Creek and the Santa Cruz River." In these areas steep-sided "cones" marked the water table as pumps sucked water from a central location, and the surrounding water cascaded into the void. Groundwater cones varied in their severity according to the efficiency of recharge and general status of the aquifer. Near Cortaro, the central part of the cone of depression descended about half a foot for every acre-foot of water removed. In the central basin, on the other hand, decline was three to four feet due to "a lack of significant nearby recharge."[18] Declining water tables had come later to the area south of Martinez Hill. Davidson supported the assertion that the old mesquite bosque and a few old cottonwoods along the river on the reservation had survived in places until declining water tables in the 1960s finally dropped below the taproot systems of the trees.[19]

Tucson had become the middle basin, and as the twentieth century came to a close, Tucson enjoyed a growing reputation for its consciousness of environmental realities, primarily because of its reliance for water on the disappeared river. In his 1998 article, Mike Davis lowered the rhetorical boom on Las Vegas, but in so doing noted, by contrast, Tucson's positive environmental ethos. Davis correctly asserted that most of the Southwest maintains "a slavish dependence upon cheap water and energy," but Tucson, "with its self-imposed environmental discipline, constitutes a regional exception."[20]

No other facet of life in Tucson serves as a better indicator of the city's environmental ethos than the "Beat the Peak" campaign, orchestrated by Tucson's water utility since 1977. Simply stated, "Beat the Peak" is a public relations program that encourages Tucsonans to conserve water during the summer months of peak water usage. It is a voluntary program that has long been noted for its success in creating a climate of environmental responsibility that makes asking for a glass of water in a restaurant a serious matter. The main focus of the program is to get Tucsonans to limit their watering of lawns and gardens to low usage hours, early in the morning or later in the evening, thus avoiding the high-demand hours during midday. Civil authorities take notice of egregious wasting of water, and police have been known to issue citations for watering a sidewalk or curb until the street runs with water. Mostly the program makes it politically correct for Tucsonans to suffer through one-flush-is-hardly-enough toilets, meager showers, and meals without water while out on the town.

Tucson exceptionalists may easily overstate the painful quality of these self-denials, however. The conservation ethos also makes it socially acceptable to landscape the front lawn with rocks, gravel, and native or near-native desert vegetation. Of course this frees the homeowner from Saturday morning lawn-mowing chores, not too much of a burden on those blistering summer mornings when the temperature is ninety-plus before the morning paper hits the front step.

But just how accurate is the description and self-image of Tucson as a regional model of "self-imposed environmental discipline"?

By the 1970s, the Tucson Water Company managed hundreds of wells and was tapping the water resources in other valleys on the fringe of the basin, but as the city grew and the water table dropped, supply and the peak capacity of the water system continued to struggle to meet demand. In this climate of almost continual crisis mentality, and as the city's slow-growth and environmentalist movements coalesced, the water politics of the city became increasingly rancorous.

The most spectacular example of water politics intruding into the electoral patterns of the community occurred in 1976, when the slow-growth movement in Tucson ran afoul of a water rate increase. Opponents of the rate hike viewed the increase as a pernicious effort to control growth by raising the cost of water. Proponents of the new rate system considered the plan commonsensical, egalitarian, and required by the ecological realities of the region. Several reform Democratic candidates advocating planned growth and "infill" (developing vacant land inside the city rather than continuing the pattern of sprawl) had won election in 1972 and 1974, including the nominal leader of the slow-growth movement, Ron Asta, who was elected to the Pima County Board of Supervisors in 1972. On the city council, reform Democrats enacted the new water-rate system in the summer of 1976. The system included lift charges that required residents at higher elevations—the foothills, for example—to pay for the increased pumping costs; new graduated rates for amounts used, establishing a rate structure that matched expense to the amount used (the more water consumed, the higher the rate); and, system-development charges that required developers to pay for the cost of expanding the system into new areas. The lift charges in particular drew attention after some customers found that water bills had quadrupled from June to July, while many others—including many non-foothills customers—had bills that doubled. The new rate system was part of an effort to recover the costs of, and hopefully limit, sprawl, but generated protest from business interests who considered the slow-growth movement a threat to local prosperity.[21]

The slow-growth movement also questioned the feasibility of CAP water. As the project's construction proceeded into the 1970s, Tucson had to make a decision whether to contract for a share of the Colorado River water. In Tucson, some supporters of the slow-growth movement questioned the CAP's cost, its reliability over the long haul, and the quality of the delivered water. Opponents of the CAP used climate studies that questioned the overall amounts of water in the Colorado River, stating flatly that the river's water was over-allocated. Opponents also used the economic analysis of the project offered by University of Arizona economists, Young and Martin. These scholars had studied the project's financial records and found that the water would be extremely costly. The reform movement in Tucson thus also ran afoul of supporters of the CAP, whether related to the issues of water rates or not.[22]

Eventually a recall effort arose aimed at the three council members who had supported the new water rates. In the meantime, Ron Asta lost his bid for reelection to the Board of Supervisors. Business interests concerned by the slow-growth movement's potential effect on the city's continued expansion had heavily backed his Republican opponent, Katie Dusenbery. Asta's defeat also stemmed from personal issues of style: Asta had become a lightning rod for the issue of planned growth through his personal advocacy and flamboyance. His defeat bode ill for the reform movement. In August 1976 the city council revoked the lift charges, in part due to the uproar over high water bills, and partly in an effort to stem the growing political opposition. Nonetheless, in early 1977 the recall effort successfully ousted the three council members. The victorious recall candidates expressed their intention to roll back rates to the 1976 "pre-increase" levels. After the election, one of the defeated council members, Doug Kennedy, remarked, "the voters had been swayed by the emotional issue of their water bills, and not by the facts surrounding the water rate increases."[23]

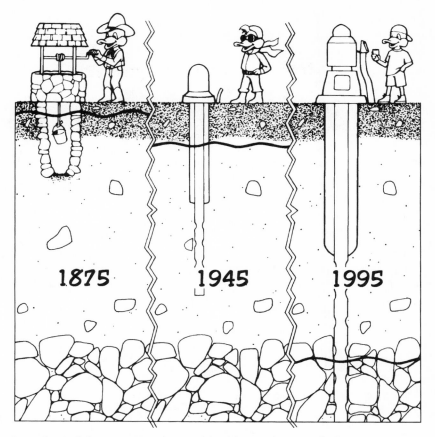

"Pete the Beak," Tucson Water's copyrighted logo and mascot for its water conservation campaigns, has exhorted Tucsonans to manage peak water demand for more than twenty years. The "Beat the Peak" campaign is credited with helping to slow the growth of peak water demand despite a rapidly increasing population. Also, the per capita use of water by Tucsonans has declined slightly since the inception of such water conservation efforts.

The recall still affects water politics in Tucson. Even the recall victors faced the same harsh realities of Tucson's water circumstance: limitations to source and system. Although the new council members did revoke the lift charges, most of the new rate system remained in place. Staffers at the water department and politicos in city hall still hesitate to bring up the issue of large rate increases, no matter how necessary, with the notion, "Remember the recall."[24]

"Beat the Peak" arose in the aftermath of the recall election. During the summer of 1977, only a few months after the recall, the water utility started the public relations campaign, including the "Pete the Beak" cartoon character, hoping to save $50 million in a projected capital improvement plan.[25] But that was not the first "Beat the Peak." As mentioned in earlier chapters, Tucsonans had grown used to shortfalls in water delivery and calls for conservation.

Water shortages were also common due to dry seasons and droughts, as well as to limitations in the water department's system of wells and mains. In the years leading up to 1920 and after, restrictions on water usage in the summers had come to be expected, and it was not until the shake-up in the water department in 1921, and subsequent repairs and additions, that the shortfalls halted for a time. In a similar vein, the water department justified the bond issue in 1949 on the basis of reports of low water pressure in neighborhoods on the urban fringe. In the 1970s, once again, a water system in need of expansion had created the political crisis, in conjunction with hot, drought-like conditions. According to the institutional memory of the water department, these dual, traditional circumstances of drought and system inadequacies had brought about the "Beat the Peak" program in 1977, and not brewing political intransigence:

> That summer [1974] was one of the driest and hottest periods on record. The City's wells were incapable of consistently meeting prolonged peak periods, resulting in local service disruptions and chronic low water pressure.
>
> The utility also had to contend with the looming and very real issue of groundwater level decline (at that time the City's sole water source) and steady population increases. It was discovered that the capitol [sic] improvement program could be reduced from $150 million to $100 million if peak demands were reduced.[26]

Unmentioned is the awareness of political limitations in the community that made saving $50 million in capital improvements necessary. Was this the source of "environmental self-discipline" in Tucson? Was it simply a pocketbook issue, when expansion of the water system had become too expensive? As Hal Rothman recently remarked to me, water in the desert Southwest has always flowed toward money. Was Beat the Peak in Tucson necessary because the money necessary to keep the water flowing had started to dwindle? Were Tucsonans simply cheap? Perhaps this was the "dwindling stream" in the 1970s—a dwindling stream of money.

Another factor bringing the assertion of Tucson's environmental consciousness into doubt is the difficulty in ascribing any single environmental ethos to the community as a whole. Tucson is fraught with contention and division—"postmodern," as I refer to it in this narrative. Certainly some Tucsonans have a keen sense of environmental consciousness, but is it appropriate to describe a majority of the community in those terms? More on this later.

Has the Beat the Peak program been successful in reducing Tucson's draw of water? In statistical terms, the program has coincided with a slight reduction in the average daily use of water by Tucsonans, and the water company gives it credit for slowing the growth of peak demand as well. "Citizens have consistently used less water over time (from an average 190 gallons per person per day from 1969–1976 to an average of 163 gallons per person a day from 1977 to present), and overall peak demand has grown very slowly, despite a quickly growing city population. This reduction has helped to slow the rate of increase in groundwater decline."[27]

	1900	1922	1965	1969–1976	1976–1996
Average water use by Tucsonans (in gallons per person per day)	266	176	175	190	163

The water utility thus succinctly provides the self-congratulatory tone that in part results in the notion of "self-imposed environmental discipline." But within the rhetoric of success lies the fundamental truth about Tucson's use of water: the decline in the water table continues unabated as more and more people migrate into the basin. The rate of increasing use and the decline in the water table have slowed, but the curve is still going in the wrong direction. City wells had pumped 1.6 billion gallons of water in 1922 to service 25,000 residents (the city wells could pump a maximum of 14 million gallons per day, which the city engineer calculated would fulfill the needs of 61,000 residents). In 1990 the pumps sucked almost 70 billion gallons from the aquifer to service over 800,000 residents.[28] Currently the water table is, in places, hundreds of feet below the surface of the basin, and the hope for Tucson's future, the 1999 "Truth About Water in Tucson," centers on CAP recharge and water reclamation, issues of supply and water-system management that have consistently occupied much of the political discourse in Tucson. Consensus on these issues remains elusive, however, because of the different perceptions of the basin's water resources voters bring with them to the polls.

Into this scenario of contested politics came the floods, sweeping down the channel, paying little heed to rancorous elections and completely ignoring basin geography as defined by geologists and historians.

The flood of October 1977 set new records in discharge on the river and showed existing floodplain assumptions to be flawed. The greatest volume of water occurred in the upper basin at Peck Canyon, but the flood through Tucson, although somewhat diminished, was still powerful enough to destroy the water-gauging station. Hydrologists had to estimate the flow through Tucson at 23,700 cubic feet per second. The peak flow at Peck Canyon was 37,300 cubic feet per second. The flood caused over $15 million in damage in Nogales, Amado, Green Valley, Sahuarita, and Tucson, with most of the damage in the upper basin occurring between Nogales and Tubac. In Nogales, washes and drainage ditches overflowed, flooding homes and an apartment complex. The river also left 3,600 acres of cultivated land awash in Santa Cruz County and flooded 12,000 acres in Pima County. It spread to a width of over a mile at Sahuarita, was 4,000 feet wide between Tucson and the village of Rillito, and spread to over four miles wide in places near Redrock in the lower basin. Occurring during an El Niño cycle of above-average rainfall, the flood had resulted from torrential rain from tropical storm Heather, which dumped rain on northern Mexico and southern Arizona from October 5 to October 10. As much as six inches of rain fell at the height of the storm in the valley, and observers noted that more than twice that amount fell on the mountains surrounding Nogales.[29]

The river channel above and below Tucson indicated differences in the stability of the riverbanks and experienced different outcomes. In the Canoa area, relatively stable flows in the 1950s and some bank "armoring" efforts had created an established channel until the flood of 1977 widened the channel primarily along those stretches not armored or channelized. In the Cortaro and Marana reaches, the river channel had been stabilized, and even narrowed somewhat with the growth of vegetation after a Tucson sewage treatment plant at Roger Road began discharging effluent into the channel in 1966. Along these reaches lined with vegetation, the 1977 flood resulted in no channel widening.[30]

Another huge flood in October 1983 served as a reminder of both nature's power and the extent of human engineering in the river valley. The continuing bank stabilization program through the Tucson Basin kept damage to a minimum, but downstream flooding inundated thousands of acres of farmland, and the river in the upper basin washed out a section of Interstate 19, temporarily severing the main transportation link between Tucson and Nogales. South of Nogales, Sonora, a bridge washed out on International Highway 15 north of Magdalena. Entrepreneurial Mexican farmers used their tractors to drag cars across the river, charging 500 pesos per haul. Near tragedy struck in Patagonia, as two people were pulled from their cars just before the swirling Sonoita Creek dragged the vehicles into the flood. And real tragedy hit Marana, when a rescue helicopter crashed, killing the pilot and paramedic on board. The helicopter had been heading north of Tucson to Catalina, which had been cut off by the floods, to pick up a "pregnant woman with complications." In Marana, the entire population of 3,500 was evacuated, and in Tucson all the bridges over the Santa Cruz River were closed. News cameras caught the startling scene as a few structures tumbled into the rushing water. Rather than terror and despair, however, the flood seemed to engender Tucson residents with a bit of postmodern bravura. As the newspaper reported, "thousands of area residents gathered at the river to cheer the rare and awesome display."[31]

Conditions leading to the flood in 1983 began with a normal monsoon season in August that turned southern Arizona green with grass and flourishing mesquite. Drainages ran in their normal wet-season fashion, and recharge of shallow aquifers, such as in the Nogales Valley, proceeded apace. The floods that followed the monsoon in late September would not have occurred if the watershed throughout the region had not been thoroughly saturated in the weeks preceding. Another El Niño weather pattern had developed, and with the river channels and aquifers as wet as they had been in recent years, a series of tropical storms surged through the area, dropping consistent and heavy rainfall throughout the Santa Cruz River watershed for five days from September 28 to October 2. The rain was so dense and consistent over the entire region that there was little variation between rainfall amounts at valley and mountain elevations, with the valley receiving almost six inches and the mountains, particularly the west flank of the Santa Rita Mountains, receiving less than eight inches.[32]

The flood down the river began on the third day of rain and by October 1 the water in the river was rising steadily. In the wee hours of October 2 peak flows coursed down the river, passing Continental at about 3:00 am with a volume of almost 45,000 cubic feet per second (a little over one acre-foot every second). At 6:00 am the crest of the flood passed Cortaro, downstream from the Rillito confluence, with a volume of 64,970 cubic feet per second (1.49 acre-feet per second). The monitoring officials estimated that a peak flow of 52,682 cubic feet per second (1.2 acre-feet per second) had surged through Tucson that morning of October 2 sometime between 3:00 and 6:00 a.m.[33]

To place these flows in context, note how long it would take an acre-foot of water to pass Congress Street in central Tucson, starting with engineer Culver's "reliable" low flow amount in 1884 to the peak flow passing through Tucson during the floods of 1915, 1977, and 1983.

1884: one acre-foot in 43 minutes
1915: one acre-foot in 2.9 seconds
1977: one acre-foot in 1.85 seconds (1.17 seconds at Peck Canyon)
1983: one acre-foot in 0.83 seconds.

The flood down the river channel in 1983 caused damage upstream from the Tucson Basin and downstream as well, but within the middle basin most of the flood was contained in the river's arroyo. Bank stabilization projects over the preceding decades, including a recent project just upstream from St. Mary's Road, had surprised officials with their durability, and the only damage had occurred where no bank stabilization had been undertaken, or in those places where soil cement on the banks abutted the native earth. In those areas, "lateral erosion" had occurred, and several homes and bridge approaches were swallowed up by the flooding river.[34] At Cortaro and Marana the 1983 flood stripped the vegetation from the banks and widened the channel from 250 feet to 450 feet at Avra Valley Road, and from 150 feet to 270 feet at Cortaro Road. In these areas vegetation quickly returned to the riverbanks, and within three years of the flood the channel at Cortaro had decreased to 170 feet due to "low-flow incision, in-channel deposition, and revegetation."[35]

In the river's reaches between Continental and the San Xavier Reservation, the rushing water had enlarged the channel and changed its course slightly. Downstream from the middle basin, the flood had overrun its shallow banks and spread out over the valley, in places by almost ten miles. One result had been the erosion of the old headcut at Greene's Canal at the site of the old Santa Cruz Reservoir Project. Since the canal's construction in 1915 and over the intervening decades, Soil Conservation Service engineers along with local farmers had constructed baffles and weirs to mitigate against further erosion at the headcut. The efforts had been largely successful, and no arroyo such as had occurred in the Tucson Basin had developed through the Santa Cruz Flats. The flow in 1983 was so great, however, that the headcut at Greene's Canal eroded upstream past the farmers' and engineer's "protective structures."[36]

The engineering efforts at Greene's Canal had proved no match for the 1983 flood, but the human artifice applied to the arroyo through the Tucson Basin had fared much better. The homes collapsing into the Santa Cruz River and Rillito Creek had made for some spectacular news footage in Tucson, but all in all the city and county officials were pleased with the ability of the stabilized channel to handle the peak flows, and set upon a course of action to further armor the riverbanks with soil cement. Local governments also ordered another round of floodplain studies so as to redefine the peak flows upon which the engineers would base their stabilization efforts. The new yardstick developed by the post–1983 studies was slightly less than 70,000 cubic feet per second through the Tucson Basin.[37]

Earlier flood-control polices had been the result of federal mandates and state legislation prior to the flood of 1977. Since the 1977 flood surpassed the previous 1915 peak-flow amounts, the flood caused a rethinking of flood-control policies and floodplain-management regulations. Individual perceptions shifted and engineers and planners came up with new parameters, but the same modern faith in an engineer's ability to control, manage, and improve nature held sway in the river valley.

Regulation of floodplain development had started in the 1970s due to federal initiatives in flood insurance and disaster relief. Congress passed flood insurance legislation in 1968 and the Flood Disaster Protection Act in 1973, which stimulated local governments in the identification and management of floodplain development. The federal grants and programs required that local governments conduct studies to determine the geographic extent of floodplains, the frequency of floods, and the volume of their flows at 10-, 50-, 100-, and 500-year magnitudes. The yardstick for floodplain regulation would be the 100-year flood zone—an unfortunate label for such political initiatives because the "year" moniker misleads the public. So-called 100-year floods can occur at any time, and do not obey any sense of chronological regularity. Although the methods of calculating 10-, 50-, 100-, and 500-year flood discharges remained debatable and often proceeded from assumptions derived from greener parts of the country (with rivers that ran consistently year-round), the federal initiatives lead to legislation in the Arizona legislature in 1973 to establish Floodplains Boards throughout the state. The next year Pima County enacted an ordinance establishing governmental management of the floodplain, and over the next several years the county shaped its building codes to include "setback requirements" for developments near the river. Eventually the regulations differentiated between property uses near "major watercourses," requiring structures to be "setback" from flood zones at varying distances determined by the structures' designed use.[38]

With evolving regulations distinctly in the background, building continued near the floodplains with a complacency that matched the city's casual overdraft of its water resource. Just as engineers and scientists had brought reassurance to Tucsonans over a virtually boundless groundwater supply and plentiful resources outside the basin, the new pronouncements regarding 100- and 500-year floods lulled the public into a postmodern sense of relativity: the river only popped into their consciousness during floods, and had nothing to do with the water from the tap or hose bib. Additionally, a recognition developed that 100-year floods have nothing to do with centuries.

The 1977 flood showed that nature could rear its head upon occasion and throw human contrivances such as highways, housing developments, and cultivated farmland into complete disarray. The flood also caused a rethinking of the recently enacted floodplain regulations. The regulations had been predicated on assumptions about flood discharge that were based on scientific models that generally assumed stable channels and constant watershed conditions. The maps of 100-year floods developed at the initiative of federal legislation in the 1970s had taken into account the magnitude of peak discharge in Tucson that remained in the range of the 1915 flood. After the 1977 flood, planners increased their estimates of discharge ranges to accommodate greater flows in the future. Confident in their new estimates, the city encouraged the development of over one thousand homes along the west bank of the river south of St. Mary's Road. The city felt safe in allowing the construction because the riverbanks in the area had been "stabilized" with soil cement "retainer walls" over eight feet thick, and the channel had been engineered to withstand a flood more than 50 percent greater in magnitude than the 1977 peak flow. Additionally, the city required that the floors of the homes be no closer to the 100-year flood elevation than one foot. The confidence of Tucson planners also exhibited itself in 1978 in the Santa Cruz River Park plan. The plan for jogging and biking trails and picnic tables along the bank overlooking the arroyo would not have made any sense had the city expected the banks to erode and tumble into the river. Given the relatively minor damage to structures in the Tucson Basin, the modern faith in engineering expertise got a big boost in 1983. But just to be safe, setback requirements were increased to 500 feet from the 100-year floodplain.[39] Not until the debacle of CAP water delivery in the early 1990s would the faith in engineering and science receive a shock sufficient to challenge the modern perception.

As an indication of the continuing faith in scientific and engineering solutions, in 1980 the Tucson water utility began a study in conjunction with the USGS of the subsidence and groundwater depletion in the city's well fields. Since that initial study, the department's hydrologists have issued an annual report detailing the decline in the aquifer at over one thousand well sites in the Tucson Basin and another 270-plus sites in Avra Valley.[40] One result of the monitoring was the clear understanding among hydrologists and some political leaders that groundwater pumping in portions of the aquifer had reached critical points, and that regulation of some sort must be achieved. In 1980 the state legislature passed the Arizona Groundwater Management Act, in part to mollify the Carter administration's assault on the CAP (to be discussed in the next chapter), and in part to reach an accommodation between major water users over future allocations of the resource. The act created Irrigation Nonexpansion Areas (INAs), which regulated new agricultural uses while allowing existing farms to persist in their current water use practices. The act also established Active Management Areas (AMAs), which sought to encourage conservation efforts and discourage increased pumping or new well drilling. One of the difficulties in groundwater management in the middle basin is the existence of several water companies and independent users outside the control of the city water utility (for example, the University of Arizona and neighborhoods such as Winterhaven, Amphitheater, and Flowing Wells). The lack of a basin-wide water authority places limits on the effectiveness of any management plan. Outside the INAs and AMAs, the act required that all new wells be registered, and attempted to limit further development of water resources. In general the act, as its name indicated, concerned groundwater and left surface flows unregulated. An exception was the Santa Cruz River. Eventually, the legislature responded to requests from Santa Cruz County, and extended the AMA to preserve what was left of the river's aboveground flow near Tubac.[41]

One result of the debate and passage of the Groundwater Management Act was the increased call by some activists and leaders in Tucson for conservation of water in the basin. Initially, the water department aimed

Beat the Peak at reducing the peak demand for water primarily through limiting lawn irrigation. In the 1980s, however, a wider conservation ethos began to spread at the behest of the water department. Reduced-flow water fixtures and drip irrigation became the norm in Tucson. As a result Tucson became known for one of the lowest per capita water consumption rates in the semiarid Southwest, and this contributed to the "environmental discipline" noted by Davis and others.

No matter how seriously, or tenuously, one takes the image of Tucson as an environmentally aware city, it is clear that the 1980s saw a rise in the awareness that water resources needed to be monitored and regulated. Whether that awareness achieved a political majority remains to be seen, but an offshoot of these efforts at monitoring and regulation was a clear increase in the effort to preserve riparian areas. The environmental movement gained momentum through the 1970s and flowered into significant restoration and preservation projects in the 1980s. One such effort resulted in the Cienega Creek Preserve, which eventually found Pima County owning about 4,000 acres on Cienega Creek, the source of Pantano Wash and Rillito Creek, and one of the last perennial flows in the region. As the urban area continued to sprawl toward the southeast, the creek had become threatened. To forestall the construction of a planned subdivision near the creek, Pima County used funds designated for flood control to purchase the preserve. County officials rationalized that construction in the creek area would have caused flooding in downstream areas already suburbanized, thus the preservation of Cienega Creek had become a watershed and flood prevention issue. The area's scenic qualities also motivated the preservation. Birdwatchers and other naturalists recognized the creek's value and pushed for the working arrangement of the preserve to allow public access on "a very limited basis."[42]

A riparian area also developed downstream from the Roger Road wastewater plant in Tucson. Pima County's discharge of treated effluent into the river channel does not match the flow in Santa Cruz County, in part because Pima County and Tucson have made efforts to expand the use of treated effluent to water golf courses, parks, and school yards. As of 1999, over 200 sites received effluent including thirteen golf courses, twenty-five parks, and thirty schools.[43]

Even in the face of increasing water consciousness, however, the urban area continued to expand with a rush and clamor that gave credence to all the old modernist assumptions. As population levels increased in the foothills surrounding Tucson, the water company faced renewed water pressure problems. The solution was to construct reservoirs at higher elevations to provide gravity flow to the growing number of homes and apartments in the hills. The expense of these system expansions also had potential ramifications for the water politics in the basin. Since water rates still included no lift charges, all water customers throughout the basin would bear the added expense to deliver water to the foothills. In other words, cheap water in the foothills is subsidized by water customers in the valley. The passionate water politics in the valley gave no sign of abating.[44]

CAP, Broken Pipes, and Groundwater Spikes

1983 to the Present

MANY IN THE VALLEY WOULD DECLARE the Santa Cruz River already dead, and in the context of the river's perennial flow through the middle basin, the declaration would be largely correct. But such a simple pronouncement ignores the perennial flows persisting along tributary streams and over stretches of the river itself in the San Rafael Valley. There are also riparian areas downstream from wastewater plants where surface flow exists. Although one might take exception that the flows of treated effluent are "artificial" and thus should not be considered part of the historic river, such a judgment would clearly emanate from a distinct perception of the river, and would not be shared by residents of the valley approaching the river with a different vision. As the record flood receded in 1983 the river returned to its former condition, a lessening stream, much changed from the river that attracted my great grandparents to the valley 120 years ago, but a stream that remained sufficiently life-giving to support the human societies in the valley.

As the sole water source, however, it seemed to be falling more and more into arrears. The panacea for the problem of source and supply arrived at last when the Central Arizona Project finally delivered Colorado River water to Tucson in 1990. But the delivery itself became notorious, and the Tucson Water Company still struggles to deal with the results. When CAP water hit Tucson homes in 1992, the valley's water politics took a lurch into the twilight zone.

At the time of its planning and construction, opponents of the CAP questioned its engineering and economic feasibility. But in its initial appearance in the valley, neither supporters nor opponents expected the fate of the project to turn on cultural perceptions and political reactions of the most grassroots nature. The Santa Cruz River had long since disappeared beneath the sand, but in 1990 a concrete-lined, uphill flowing river returned to the middle basin. The water in the canal, in the aquifer, or flowing out of the sewage treatment plant at Roger Road, expressed no preferences or emotion as to its point of origin or flowing circumstance. The people in the valley provided the preferences and passions.

Tucson is the terminus for the CAP system, the last of the big reclamation projects that developed water in the semiarid West. The canals deliver water 336 miles from Lake Havasu on the Colorado River, traveling through central Arizona to Phoenix before turning south to Tucson. To lift the water 2,400 feet from the Colorado River Valley to the Sonoran Desert outside Tucson requires fourteen pumping plants (the actual lift is 2,900 feet, since the water flows downhill in places, and has to be lifted again to regain the lost elevation). Surrealistically, the canal briefly follows the course of the Santa Cruz River as it enters the middle basin, flowing in the opposite direction of the old river. The straight lines and angles of the canal speak of the system's modern origins, a brainchild of engineers and scientists, but the overall concept of the project is definitely postmodern. The logic of the system exists only within narrow boundaries. Critics of the system also speak from within their own parameters of logic, creating a discourse surrounding the CAP that is often chaotic.[1]

The CAP had its origins at the turn of the century in the appeals of visionaries, or, depending on your perspective, the crackpot ravings of lunatics. George Maxwell, the gadabout and bombastic booster of reclamation in the West touted the "Highline Canal" to Arizona Governor Hunt in the 1920s. Engineering reports in 1921 and 1922 conducted by the USGS and University of Arizona noted that the Highline Canal, whether starting at Boulder Canyon, site of the proposed Hoover Dam, or at Glen Canyon, also site of a future Bureau of Reclamation dam, would be difficult, though possible, to build. While the canal would allow irrigation of thousands of acres of land in the central valley, both studies remarked that its expense would make it problematic. In the words of the USGS and Reclamation Service report, the canal would be an "extravagant absurdity."[2]

There was also the problem of water availability. How could Arizona siphon off enough Colorado River water to fill the canal and irrigate the fields when there were six other states, not to mention Mexico, laying claim to a share of the river? To settle the issue of competing claims for the river's water, Secretary of Commerce Herbert Hoover called representatives of the seven Colorado River Basin states (Arizona, California,

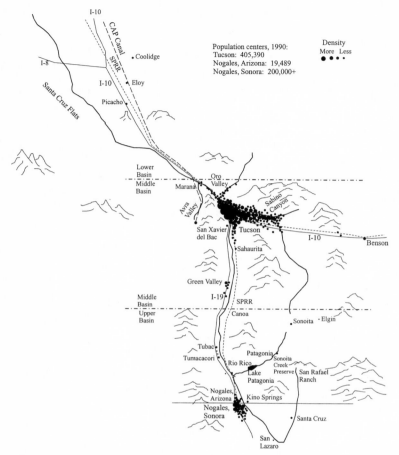

As the population in the river valley continued to increase throughout the last decades of the twentieth century, new settlements arose on the sites of old towns and villages. Green Valley developed as a retirement community within sight of the agricultural settlement of Continental. Rio Rico occupied the bluffs overlooking Sonoita Creek where the Gándara hacienda and Calabasas had once been. While communities such as Eloy, Coolidge, Patagonia, and Santa Cruz languished off the beaten path of postindustrial development, the urban centers in the valley—Tucson and Nogales—reached new levels of population and expansion.

Colorado, Nevada, New Mexico, Utah, and Wyoming) to Santa Fe in 1922. The delegates debated for months, and finally on November 25, 1922, the representatives all signed the Colorado River Compact. The agreement divided the river between upper basin states and lower basin states. Each basin would receive 7.5 million acre-feet per year. The compact established the dividing line for the basins at Lee's Ferry, Arizona, and so of the seven states, Arizona was the only one that straddled both basins. But most of Arizona's water would come from the lower basin allotment, and this put the state in direct competition with California. Whereas Arizona's need for the water remained mostly in the future, California claimed to need the water immediately, and their proposals, in conjunction with the Reclamation Service, for a huge dam at Boulder Canyon and an All American Canal to service the Imperial Valley far to the south, made those claims quite real. The Arizona legislature, at the urging of Governor Hunt, refused to ratify the compact, making Arizona the only state of the seven to fail to ratify the agreement at the time (the legislature finally voted to sign the agreement in 1944).[3]

Although the effort to distribute Colorado River water among the basin states did not seem at first glance to further the cause of the Highline Canal, the notion got a tangential boost in the 1930s when federal largess built Hoover Dam. Another aspect of the legislation for the dam was the All American Canal in the south. The canal would deliver water to the farmers in California's Imperial Valley through an unlined canal (wasting thousands of acre-feet of water to seepage) that would travel directly over the sand dunes west of Yuma, rather than dipping into the firmer terrain in Mexico. Whereas Hoover Dam was grandly spectacular and made all water projects seem doable in this age of reclamation extravagance, the canal spoke more directly to the aspirations of Arizona visionaries (or crackpots). As California received more and more federally subsidized water, Arizona leaders clamored for their own dams and canals. The construction of Parker Dam and another aqueduct in the 1940s delivering more Colorado River water to southern California, primarily Los Angeles, further cemented Arizonan's suspicion for its neighbor to the west, and fostered development of the idea for a Central Arizona Project.[4]

Eventually taking the lead from ranting state governors, Arizona's Democratic Senator Carl Hayden struggled consistently through his decades in Congress to put together the political support necessary to secure the project. Hayden had to maneuver around California's resistance to any plan that might diminish their draft of water, as well as the resistance of upper basin states with their own interests and plans. In addition to the political opposition, nature seemed to be obstructing the Arizona canal as well. By the 1960s it had become clear that the Colorado River's flow had been over-allocated. The settlement in 1922 had been based on overly optimistic projections of the river's flow, and so the distribution of water to the upper and lower basins, and to Mexico, was in reality about five million acre-feet more than the river typically contained. The upper basin states, especially Colorado, were aware that if Arizona joined California in using all of their share of water on the basis of the over-allocation, there would be no water left for the upper basin states for any future development efforts. The correction in the projections of the river's flow only heightened the political difficulty of Hayden's task. The political maneuvering took decades. Hayden was ninety-one years old and routinely falling asleep during Senate deliberations when President Lyndon Johnson finally signed the legislation for the Central Arizona Project in 1968. The main reason why Hayden stayed in the Senate so long, beyond the point of embarrassment, was to shepherd the legislation through Congress during the waning days of Johnson's second New Deal, the Great Society.[5]

A product of the reclamation philosophy that dominated the western region for at least eighty years, the CAP soon ran afoul of 1970s political and economic realities. President Carter sought to bring his own version of logic and commonsense to such "pork barrel" projects as the CAP, and threatened to cut funding for this and other western water projects. The dispute escalated into the so-called Sagebrush Rebellion, wherein western politicians waged a rhetorical battle with the president and his cabinet. President Carter complained that the western water projects made little sense unless coupled with a conservation ethic, and so required that at least some modicum of planning and regulation of water utilization be adopted by those states receiving the massive federal subsidy for water reclamation projects. At least partly as a result of this political squabble, the Arizona legislature passed the Groundwater Management Act in 1980. The law identified zones in the state where water was scarce and required regulation, and where limitations on further development was perhaps necessary. The act established one such zone in the Tucson area, and so the squabble over CAP funding resulted also in an effort to legislate wiser water use.[6] One of the things the legislation could not accomplish, however, was to reduce the cost of building the project or of the delivered water. As critics had pointed out in the planning stages, the cost of pumping the water to such an elevation and over so great a distance would result in charges that could bankrupt farmers, and it was questionable whether even the richest urban areas could afford it. As to mines and other industrial customers, the harsher qualities of CAP water made its applicability questionable. Whereas CAP water actually proved to be less alkaline than some groundwater, the

Colorado River water in the canal was thought to be unsuitable for use in the copper refining process. As a result, when the canal reached completion in 1990, few farmers or mines had contracted for the water, and Tucson remained one of the only customers for the CAP in Pima County.[7]

The story of the CAP's arrival in Tucson homes has achieved mythic proportions in the brief time since the debacle unfolded. Tucson Water mounted a public relations campaign prior to the introduction of CAP water in November 1992. The utility trumpeted taste tests, but also warned customers that CAP water might kill fish in aquariums and pose a threat to those with certain health problems. It seems incredible now that the utility and Tucson's political leaders could not have foreseen the problems with the CAP's introduction, but perhaps the political myopia of 1992 indicates the power of that modern panacea. The CAP had been part of the water consciousness of the city for almost a century, and as such had acquired an inertia that was unstoppable.

CAP water, traveling hundreds of miles uncovered through the desert where evaporation would tend to concentrate the solids and minerals in the water, was far harsher than Tucson's normal groundwater supply delivered from the aquifer under the basin. As such it had to be treated, and the water company constructed a plant in Tucson to render CAP water potable. Even after treatment, however, Tucson's CAP water leached rust out of old galvanized pipes, turning tap water a dirty orange color. In the worst cases, pipes simply melted away. About 84,000 customers received CAP water initially, or about 58 percent of the water utility's system. The political fallout from these residential disasters has largely determined the uses, or lack of use, of CAP water to this day. The water company tried inexpertly to manipulate the chemistry of the water to alleviate the problems, but the damage to pipes and the utility's credibility had already been done (it should be noted that other municipalities, such as Phoenix, had no such problems with CAP water). After about two years, deliveries of CAP water in Tucson stopped. By 1995 the utility had paid out over $1 million for property damage, while steadfastly denying any responsibility. In the political fallout, the head of the water utility resigned, as did the CAP plant manager. The long-term damage to the water company's public image is untold. Other reactions to the introduction of CAP water included an increase in the use of bottled water and an increase in the use of in-home water treatment systems.[8]

Another casualty of Tucson's CAP experience was the loss of consensus behind the modern faith in the engineers' ability to solve the basin's water dilemma. The CAP had been the silver bullet solution for decades, regardless of whether that expectation was fair or accurate. In part, politicians had created support for the expensive project by over-selling it, and then when the water arrived, it undercut the modern perspective of water and the river, and served as a clear indication of the presence of what I refer to as the postmodern perspective of water in the valley: what flows out of the taps and hose bibs is not the Santa Cruz River. If not CAP water, it is invisible or distant groundwater, or bottled water that you buy in a grocery store. For most Tucsonans, thoughts about the origins of water are abstract, akin to the conceptual distance between beef and chicken in the barnyard and the cellophane-wrapped conveniences found at Safeway. My point is not to condescend or criticize Tucsonans for some basic ineptitude or ignorance, only to point out that most residents in the valley have a postmodern, disassociated, abstract perception of the valley's water supply. The residents of Tucson still rely on the Santa Cruz River and its aquifer, but the river and its aquifer make little impression on their consciousness. The electoral politics surrounding the CAP give clear indication of this emerging abstract perception.

By 1995 opposition to domestic use of CAP water had coalesced behind the Water Consumer Protection Act (WCPA). The act, placed on the No- vember 1995 ballot as a result of petition signatures, required, among other things, that CAP water match the quality of groundwater before it could be delivered to homes. Failing that quality—virtually impossible to achieve given the nature of the CAP source and its long travels—the act basically outlawed use of CAP in Tucson homes. A reverse initiative appeared in 1997, asking voters to repeal the WCPA. When voters defeated the initiative, the utility seemed forced to shift its focus to artificial recharge and other non- domestic uses. Despite CAP's defeat at the polls, however, the water utility and city council persisted in their efforts to use CAP water residentially. I will summarize this postmodern electoral saga later in this chapter.

CAP's effect on the basin's water politics was clear and startling, but what was its effect on the river and its aquifer? For a brief time delivery of CAP water to residential users allowed Tucson Water to stop pumping from sev- eral of the city's wells. Hydrologists had been monitoring the water level in the wells for decades, marking the steady decline in the water table. When the pumping ceased, hydrologists noticed an almost immediate reversal in the water table charts. Water tables began to rise with impressive and dramatic rapidity: a groundwater spike. As deliveries ceased and pumping resumed, the old pattern of overdraft returned and the graphs tracing the decline in the water table returned to a depressing downward vector.[9]

The failure of the CAP put the modern perspective under assault, but the faith in scientific and engineering solutions was not vanquished. The modern perception persists, but its following is much circumscribed, and the prophets of that vision more often than not preach to the converted.

An indication of modernism's persistence is the latest in the long line of studies of the water supply that appeared in 1999, published by the Water Resources Research Center at the University of Arizona. In the tradition established by G.E.P. Smith, the University of Arizona researchers delineated the basin's water situation while trying to steer clear of the water politics that determine the uses of water in the area: "no direct solutions are recommended."[10] The study, "Seeking Sustainability," provides much useful information, but dodges the central fact about the city's use of water. Any reasonable scenario of effluent usage, conservation practices, and transitions from industrial and agricultural uses to domestic use, will continue to result in overdraft of the aquifer. There are simply too many people in the basin to achieve balance between draft and recharge. The study does clearly state that subsidence problems will limit the city's ability to tap groundwater indefinitely, since drawing water at such rates has the effect of turning newer alluvium into older alluvium: without the water saturation, the ground compacts into such dense formations that recharge of the aquifer and wells slows to a crawl. The productivity of wells then plummets.[11] This has already taken place in some of Tucson's wells, where output from the wells in 1999 had declined by a third or even a half since the late 1950s.[12]

Studies such as "Seeking Sustainability" contain all the information necessary to fuel a hopelessly tangled debate over water resources in the basin. On one pole are the naysayers seeking validation for the position that there really is no problem worth worrying about. They can point to the simple assertion that the aquifer remains 89 to 91 percent unused; as the report stated: "9 to 11 percent of the total has been withdrawn."[13] On the other pole are the doomsayers, pointing to the occurrence of subsidence and declining well output to make the equally true assertion that the remaining water in the aquifer will be inaccessible in the future, given the results of huge overdrafts now. Occupying a middle position between the two poles are the authors of "Seeking Sustainability," and many other cautiously optimistic Tucsonans, expressing the hope that an informed public (informed by the expert knowledge of scientists and engineers) and rational democratic processes will create the formulation and implementation of a viable water policy. The basic conundrum remains, however, that too many people in the valley will thwart any reasonable strategy. The end will come to Tucson's use of groundwater, sooner or later. So how long can the water last? According to the 1950s "Truth About Water in Tucson," the city has already reached the end of "reasonable pumping lifts." But as it turned out, the engineers and scientists in the 1950s seem to have gotten it wrong. Most of the wells are still flowing, and the pumping lifts still seem to be reasonable enough. On the other hand, an erroneous projection by scientists and engineers fifty years ago might reasonably be expected to diminish the hope of those relying on scientific and engineering projections today for the rational formulation of water policy and the answer to the question, how long can the water last?

Away from Tucson and its convoluted water politics, development continued in the upper basin along an established trajectory. The ranchers in the San Rafael Valley continued to graze livestock in generally modest numbers and the little stream continued to meander out of the Canelo Hills, through the valley and into Mexico. Access to the valley was still limited to dirt roads, and so the wave of development that came to dominate other portions of the valley due to the railroad and the interstate highway largely bypassed the high grasslands. The small communities of Lochiel and Santa Cruz seemed to be snoozing through the twentieth century. But a hint of change seemed to be in the air. In the San Rafael Valley, owners of the 1,800-acre Ki-He-Kah Ranch in 1993 sold off portions of land in 160-acre parcels. The new owners of the lots were then legally entitled under Arizona law to subdivide the 160-acre parcels into 2.5-acre lots. The unwelcome prospect of low-density residential development had arrived in the high valley at last. Other factors that now threatened continued ranching in the valley were estate tax laws. As long-time ranchers grew older, children and grandchildren had to contemplate paying estate taxes that could run as high as 55 percent, and hundreds of thousands of dollars. The tax laws and economic pressures of the late-twentieth century seemed to be leading to the break-up of the large land holdings in the valley. Whether this would actually happen, and what effect such a development in land holding would have on the river, remained to be seen.[14]

The Nogales, Arizona, population in 1997 was about 20,000, and its sister city across the border in Mexico about ten times that size. The official population of Nogales, Sonora, was 180,000, with perhaps another 20,000 "floating" inhabitants seeking work in the factories along the border. Although agricultural and mining demand remained limited in the vicinity of the border, the expanding maquiladoras factory system straddling the border increased water consumption (and pollution concerns) in Mexico. The municipal system of Nogales, Sonora, continued to serve only about half of the residents of the city, with the sewage and runoff from this system flowing down Nogales Wash into the international wastewater treatment plant, where the treated effluent then produced the newly instituted perennial flow of the river at Tumacacori and Tubac. In addition to the sewage from Mexico within the system, a stream of tainted water flowed down Nogales Wash through the concrete tunnels and culverts under the border. This perennial stream conjured up no enthusiasm or romantic celebrity, such as the riparian area just downstream enjoyed. As a state water biologist recently noted of the flow in Nogales Wash: "You don't want to drink it or play in it. The smell would drive you away!"[15]

The flow to celebrate was downstream from the treatment plant. Threatening the renewed riparian area was the possibility that Nogales, Sonora, might develop the ability to treat their own sewage and therefore discontinue the practice of piping it into the U.S. treatment plant. Part of that perennial flow was furnished by water imported into Nogales, Sonora, from a well field in the Magdalena watershed. This water from outside the Santa Cruz River watershed was piped into the international wastewater treatment plant after servicing domestic and industrial uses in Nogales, Sonora. Clearly, the CAP was not the only system to deliver water to the valley from beyond the watershed.[16]

In Nogales, Arizona, the municipal system grew to the extent the shallow and narrow aquifer would allow. As more wells and pumps in the river channel came on-line, some of the original pumps in Nogales Wash were taken out of service. Water still quickly recharged the aquifer as in eons past; the problem now was pollution. A slowly moving plume of TCE is moving north out of Mexico. It first showed up in the Memorial Park well (used to water the park and stadium fields), and so that well was shut down. There was also pollution entering the aquifer from manufacturing plants on the Arizona side of the border, and it is only a matter of time before more wells show signs of contamination. Another threat to the water system came from floods, which damaged wells in the river channel. In part because of this more-or-less constant threat, and in part out of a need to secure ever-increasing supplies of water, the city water department has gone on a search for distant water supplies. Presuming that the CAP would be long in coming, or would never reach Nogales at all, the city sold its share of CAP water and used the revenue from that sale to purchase other water supplies closer to home. The city also embarked on a maintenance and efficiency drive. In 1988, 28 percent of the water in the city system was lost to seepage and leaks, categorized as "lost and unaccounted." Much of this water leaked out of old pipes in service since 1912. In particular, the original ten-inch steel pipe up Proto Canyon from the pump house at the river was leaking so much that a perennial stream trickled down the canyon to the river. Once the pipeline was replaced, the canyon dried up again. Although some sections of the old pipe remain in the system, the lost and unaccounted percentage had dropped to 10 percent in 1999.[17]

The basic ecological reality that dominates the border cities' water supply remains the physical characteristics of the narrow and shallow aquifer. The cities have always had to cope with the limited aquifer, but they have also enjoyed the effects of quick and total recharge of the water table. This became evident most recently during the 1999 summer monsoon season. The preceding winter had been very dry and the wells in the Santa Cruz River and tributary canyons started to diminish in output. Then the summer rains came in abundance, delivering eighteen inches of rain to the upper basin, and the aquifer quickly recovered to its full potential. Without that bountiful monsoon, Nogales, Arizona, probably would have instituted a water-rationing program, and may have to yet, if future monsoons fail to deliver adequate rain.[18]

Downstream from the international wastewater treatment plant north of Nogales, the effluent flow re-created the perennial stream in the river channel, complete with flourishing cottonwood trees and marshes inhabited by ducks and geese. At its initial point, however, the output from the plant is somewhat less inviting. One and one-half miles downstream from the plant, the water contains "high levels of ammonia and parasites, including Giardia."[19] But the river cleanses itself, and the minnows and frogs farther downstream show fewer signs of disease. Since the flow of treated effluent began in the 1950s, the riparian area has taken on a permanent feel. Threats to its continued existence within the United States, primarily from private development initiatives, resulted in a grassroots effort to preserve the area's "natural" beauty. The Friends of the Santa Cruz River is a non-profit organization in Tubac that coordinates efforts by private citizens and government agencies to maintain the area's scenic quality, including a state park at Tubac, a national monument at the restored mission at Tumacacori, and a four-mile walking trail along the river. Volunteers from the Friends also test the water on behalf of state agencies.[20]

With the flow in the upper basin somewhat stabilized in the San Rafael Valley by conservation-minded ranching practices, and the reestablished riparian area downstream from the international wastewater treatment plant, the river probably comes closest to resembling its preindustrial configuration in the upper basin. In fact, so much water now flows in the channel near Tubac, it may surpass any perennial flow within historic memory. It perhaps resembles the lush environment first witnessed by human eyes at the end of the last ice age, before the current semiarid climate took root. If the observer can ignore the urban area filling Nogales Wash and sprawling up the draws and hollows, houses dotting the hillsides at Rio Rico, telephone poles and satellite dishes, and semis droning on the interstate highway, the sight of the green, flourishing riparian area surrounding Tumacacori National Monument and Tubac Country Club must look something like it did, if not one hundred years ago, then ten thousand years ago.

When the CAP arrived in the lower basin in the 1990s, doubts about its financial viability quickly surfaced. Irrigation districts did go bankrupt, and others simply refused to sign contracts for CAP water due to its high costs. In order to recoup the costs of constructing the canal system, early charges for the water topped $50 an acre-foot. One example of lower basin farmers facing the CAP dilemma was the Central Arizona Irrigation District in Eloy, Arizona. The district contracted for CAP water, but soon found itself unable to meet its payment obligations. At fault was the high cost of CAP water, but also declining commodity prices put district farmers into an impossible situation. In the ensuing reorganization, completed in 1996, the district maintained its access to 132,000 acre-feet per year of CAP water, a circumstance that will remain in effect until 2003 (the district's CAP water is in two "pools": about 72,000 acre-feet of original contracted water at $23 per acre-foot (the readjusted price) and 60,000 acre-feet of "relinquished" water at $33 per acre-foot). After 2003, the district will fall under a water settlement currently being negotiated between state officials, the federal government, and several Native American tribes. The district expects to receive somewhat less "excess" CAP water at that point—about 112,000 acre-feet per year. That water will be distributed to farmers occupying the 87,000 acres of the district, although generally no more than 60,000 acres is under cultivation at any given time. And of course, any CAP water going to the crops in the lower basin relieves the aquifer and groundwater supply from that task of irrigation. In 1999 the district pumped 86,698 acre-feet from the aquifer at an approximate cost of $21 per acre-foot (approximate because the cost is tied to varying electric rates). The water table has actually risen in areas of the district where the pumps have been shut down as CAP water became available.[21]

The Hohokam Irrigation District in Coolidge, Arizona, was in a somewhat different situation when the CAP price for water came down in 1993. Not having a contract for CAP water, the district acquired "excess" water not under contract to other agricultural, municipal, or industrial users. The amount available has run between 40,000 to 60,000 acre-feet per year, with the price drifting upward gradually. Last year the district used 62,000 acre-feet at a price of $27 an acre-foot (the cost of groundwater was about $25 per acre-foot). The district expects the cost of CAP water after 2003 to be $28 per acre-foot.[22]

The small communities in the lower basin, primarily Eloy and Coolidge, have served the lower basin's agricultural interests, and so remained diminutive compared to the growing metropolises to the north and south. While Phoenix and Tucson were quadrupling and quintupling in population, and expanding their boundaries by the mile, Eloy and Coolidge were growing by inches. In 1950, Eloy's population had been 3,580, growing to 7,211 in 1990. Coolidge started at 4,306 in 1950, increasing at a slower rate than Eloy, to a population of 6,927 in 1990. Coolidge actually lost population in the 1960s, something of an anomaly in the Sunbelt region.[23]

The lower basin did not completely escape the effects of Sunbelt migrations and tourism, however. The bustle of Interstate 10 created a thin ribbon of commerce in the form of gas stations, fast food restaurants, and motels. But just off the interstate agricultural fields continued to sprout in the sun, a green pastoral patchwork amidst a sea of brown, undulating desert.

Politics dominates the relationship between the river and human society in the 1990s. No aspect of the river's journey from start to finish, aboveground or below, is free from the incursion of water politics. In the latest round of the CAP controversy, a campaign poster proclaimed, "Get Politics Out of Our Water." A specious call, the notion is without a doubt impossible. No matter what stance one takes on the CAP—artificial recharge, preserving riparian areas, or regulating land-use options—politics determines the realm of possible outcomes. The politics remain largely intractable due to the differing perceptions and visions of the river held by the participants in the political tussles.

In the San Rafael Valley, Bob Sharp and other ranchers tried to use political agreements to preserve the valley's ecosystem and their own life-styles. As fear mounted that absentee landowners would begin subdividing the valley's large ranches, Sharp and other ranchers formed the San Rafael Valley Land Trust in 1994 and started searching for a strategy and mechanism to preserve the large land holdings. Sharp and his sisters owned the San Rafael Ranch, 22,000 acres, thirty-four square miles, in the heart of the valley. The land trust negotiated with the Nature Conservancy, and also contacted the state government's Heritage Fund, administered by the Arizona State Parks Board. The advantage of the state deal to valley land-owners was that they would continue to own their land, and would only sell the "development rights" to the state, which would establish a "conservation easement" and pay the Sharps $8 million out of the Heritage Fund. In return for selling the development rights, ranching activities in the valley would be protected and the Sharps would receive breaks on property and estate taxes. In return for the $8 million, the state would acquire the right to build a small "grassland education center" on the Sharp's ranch, and the area would be preserved as open space.[24]

Opposition to the plan arose from environmental groups, like the Sierra Club, which disputed the wisdom of maintaining ranching activity on the grasslands and the legality of conservation easements using Heritage Fund revenues, which come from the sale of Arizona lottery tickets. Other opposition arose from the Western Gamebird Alliance, representing bird hunters, who also criticized the easements, especially because the parks board had no land management plan for the area that would allow game-bird hunting.[25]

After all the political wrangling, the Sharps finally sold their ranch to the Nature Conservancy in 1998, and negotiations over the conservation easement continued between the conservancy and the state parks board. In January 1999, the parks board announced the agreement to spend $8.6 million to create a conservation easement in the valley. The finalized deal gave the state ownership of 3,600 acres, including the headwaters of the Santa Cruz River, as well as development rights to the rest of the San Rafael Ranch. The conservancy maintains ownership of the remainder of the ranch, and announced that it was negotiating with "prospective buyers" who would agree to maintain the open space according to the easement's stipulations. Part of the state's plan was meant to regulate the traffic through the riparian area along the river. The parks board envisioned "guided walking tours through the headwaters of the Santa Cruz River eight times a year, and only with advance notice."[26]

No matter how one views the Sharp's or the Nature Conservancy's strategy for preserving the ranch, it is clear that the solution was derived through a political process. Whether politics led to a Nature Conservancy preserve, a free-market development splurge, a ban on cattle such as advocated by the Center for Biological Diversity, or a tax-break compromise benefitting landowners, the river and its watershed were destined to be affected by the political process.

The stretch of the river through Mexico is no less subject to political forces. Water politics even intrude into the sleepy community of Santa Cruz. The community of 1,900 residents remains isolated despite the railroad connection to Nogales, Sonora. The train comes through Santa Cruz once a day but often does not even stop. The dirt road from Santa Cruz to Nogales remains a 90-minute chore, and any hope of a paved road depends on federal revenue that is unlikely to come. Even the agricultural pursuits of the village are subject to political developments beyond their control. The flourishing apple orchards near Santa Cruz recently suffered a blow from tightening tariff protection for U.S. apple-growers. On the other hand, the community's isolation doesn't hinder local activism by its residents. Local farmers in 1991 protested against the construction of a sewage treatment pond, arguing that the pond would create downstream pollution. The protest was successful and the pond was never constructed.[27]

Another political scenario affecting the river and ranching recently played itself out downstream near Nogales. As mentioned previously, with the controversy surrounding the completion of the CAP in southern Arizona, officials in Nogales decided to sell their share of CAP water. The Nogales officials felt the CAP canal would never reach the city, and even if it did, the water likely would be unsuitable for domestic use, so they sold their allotment to Scottsdale, a suburb of Phoenix. While some water department officials disagreed with the decision, the sale was completed, and the city began searching for another source of water to replace the CAP. The solution appeared in a manner adopted by Tucson in the 1950s and countless other cities throughout the West—buying up agricultural land for its water rights. In this case, the City of Nogales purchased the Guevavi Ranch, near the old mission site. The city soon began sinking wells on the ranch and constructing a four-mile pipeline to connect the wells to the old pumping plant at Proto Canyon.[28]

The upper basin traditionally extended to Canoa, where the river's flow ceased and the transition to the middle basin began. But the ranch had found itself falling into the political orbit of Tucson long ago, and nothing more clearly represents the political extension of the middle basin thirty miles to the south than the development plan for Canoa announced in 1996. Fairfield Homes announced a plan to develop 3,186 acres of the ranch south of Green Valley with over 6,000 homes, two golf courses, hotels and shopping centers, and a private airstrip. The plan generated loud debate among Green Valley residents, and members of the Pima County Planning and Zoning Commission and Board of Supervisors. Finally in January 1999, the supervisors voted 4 to 1 to defeat the plan by denying Fairfield's rezoning request. Opponents of the plan heralded the decision as the first major denial by the supervisors of a developer's rezoning request since 1973.[29] Had the political winds in Tucson fundamentally changed? Had the planned, or slow-growth, movement in Tucson been given a new lease on life? In April 1999 a citizens group calling itself the Amigos de Canoa formed in an effort to save the ranch from development by buying it. The "coalition of environmentalists, Green Valley residents, historic preservationists, astronomers, archaeological and cultural interest group members" seem to feel that the political winds are uncertain. Fairfield Homes, or some other developer, is free to submit a revised plan to develop the ranch, and who can say what the vote will be with some future Board of Supervisors? Of course raising the money to buy the ranch is another issue. As the Nature Conservancy can well attest, discussions with property owners over sale price can easily lead to long and arduous negotiations. Initially the "amigos" focused on the ranch's $10 million appraised value, although Fairfield Homes stated flatly that $10 million was too low.[30] The political winds will continue to swirl around Canoa for years to come. Who can say what the ranch's fate ultimately will be, and what effect that will spell for the river and its aquifer?

In the middle basin, of course, the CAP dominates water politics. Elections in 1995, 1997, and 1999 provided Tucsonans with the opportunity to vote on the CAP. The votes basically ran: 1995, don't use it; 1997, don't use it; and, 1999, find some way to use it. Of course the issues were much more complex than this rendering, and I will come back to the issue of the CAP referendums, but whatever policy is determined, the river and its aquifer will rise or fall according to the political determinations. Will there be more groundwater spikes, or will the water table continue to drop? Political outcomes will decide.

And of course, groundwater itself has been placed under political control, at least since 1980 and the Groundwater Management Act, and maybe beginning in 1948 with the designations of "critical groundwater areas." But as with all political manifestations, the effects of the politics are varied and debatable. Enforcement is key, and the level of enforcement varies with the political winds. As a case in point, consider the discussion of enforcement strategies employed in the Tucson Active Management Area, as part of the management plans administered by the Arizona Department of Water Resources. The Tucson AMA is now under its "third management plan," and at least according to the stated goals, the screws are being tightened:

> During the first and second management periods, the Department took a non-traditional approach to enforcement. Given the recent introduction of the Code and management plans, a high level of tolerance was employed. . . . Fines were set at low levels and probationary provisions and advisory notices were widely use.
>
> The results of this enforcement strategy have been mixed. Some mitigation programs developed under this approach have been successful in increasing water use efficiency, while others have been less effective. In most cases, significant and sometimes disproportionate amounts of time and resources have been invested by both the violators and the Department.
>
> The Third Management Plan approach to enforcement will exercise flexibility on a more limited scale.[31]

Is it correct to consider a lessening of flexibility as a tightening of the screws? Obviously the enforcement strategies exist within a political landscape. To the most extreme defenders of groundwater supplies, no flexibility would be tolerable, and violators of the code would be clapped in irons and thrown in a dungeon, but that enforcement strategy clearly falls outside the boundaries of the existing political landscape. Whether through bureaucratic initiatives or referendum elections, the river and its aquifer reside within a landscape largely determined by political influences.

As previously mentioned, water politics affecting the river are so contentious because at least three visions of the river coexist in the valley—visions we have explored in some detail. The archaic vision, referring to the reverent, spiritual, and even mystical view of water in the river, is still alive and well. Daniel Preston, vice-president of the Tohono O'odham nation, gave a clear expression of this perception at the meeting of the American Society for Environmental History in Tucson in April 1999, speaking eloquently of water's central position within O'odham cosmology. But even this vision of water must pay heed to the political landscape, as Preston and the Tohono O'odham acknowledge through their continuing efforts in the federal and state courts to gain their share of the region's water resources.

The archaic vision also motivates preservation efforts that run the gamut from the Friends of the Santa Cruz in Tubac, the Nature Conservancy on Sonoita Creek and in the San Rafael Valley, and the local, state, and federal efforts behind the preservation of the Cienega Creek riparian area. All of these efforts have at their heart the perception of water as life-giving, supporting countless species in a wondrously variegated ecosystem.

The modern vision of the river, so dominant in the late-nineteenth century, is also alive and well at the beginning of the twenty-first century. Many residents of the valley continue to view any water flowing unused or sitting in the aquifer untapped as wasted. Water in the aquifer is there for the taking, with the only criteria applied to its use being scientific expertise. Calls for conservation often come from this perception, not so much a call for wise-use out of reverence for the resource, but a conservation ethos derived from progressivism's faith in efficient management of resources. This vision of water recently surfaced in two separate proposals for the creation of recreational lakes in the valley, with the argument that water was available for the lakes either in the aquifer or through the CAP canal, and if not used for the lakes, or for agricultural or urban uses, the water would be wasted. The proposal for the lake in Pinal County—not far from the old Santa Cruz Reservoir Project—called for purchasing a billion gallons of water a year from the CAP to maintain a 400-acre lake. Without the constant resupply, the lake would quickly dry up. In Pima County, Sahuarita's town council approved a proposal for a much smaller ten-acre lake that would serve as a town park. The difference in this proposal is that the water for the lake would come from wells tapping the aquifer. Once again evaporation rates in the semiarid climate would require constant resupply. The modernist calculus came into play in this last scenario, as the proponents of the Sahuarita lake offered the rationale that the water could go for the ten-acre lake, or for: "2.4 golf holes, 10 acres of pecans, 150 homes, or 12.6 acres of grass turf."[32]

Another aspect of the modern perception of the area's water resources appears in the continuing publication of scientific reports. With a legacy dating back to G.E.P. Smith's boosterish support for dams and reservoirs, the University of Arizona and USGS continue to study and analyze the river valley's scarce water resources. As discussed earlier, the latest rendition is "Water in the Tucson Area: Seeking Sustainability," which the University of Arizona spent $100,000 to produce.[33] It is an admirable compilation of data that asserts to be objective, not taking sides in the volatile political atmosphere surrounding the question of the CAP's use. Within the report, all the conundrums and dilemmas of the city's water situation come to light, but a persistent progressive faith in logical political outcomes dominates the study. The assumption is that rational decisions about water policy can be made once the electorate has a thorough understanding of the scientific and engineering realities facing the city. The report even acknowledges that voters will take stands on these issues from their own calculus of lifestyle and personal comfort. Not only does the report express a faith in democratic process emanating from individual choice (no "mass society" skepticism intrudes), the scientists view the river and aquifer as absolutely manageable. Questions of wise use are only imperative due to the resource's limited nature. But it is a limit whose urgency is couched in rationalism. We have time to educate voters, formulate plans, and strategize our way into a balanced, renewable water regimen.

The chaotic and surrealistic postmodern vision of the river appears side-by-side with the other perceptions. The surrealism of the CAP will be complete if the plans and efforts to establish the Rio Nuevo in downtown Tucson are successful. The evolving plan is percolating through city hall and among community and neighborhood activists. The idea is to develop the river along the lines of San Antonio's River Walk, but in an indigenous Old Pueblo style, using effluent or CAP water to recreate a perennial flow through Tucson's downtown. What now appears as something of an urban wasteland would someday take on the mantle of a tourist mecca. Regardless of the plan's merits as a development scheme, the prospect of Colorado River water discharged into the engineered river channel for the sake of tourists, joggers, and shoppers, will surely bring a smile to the face of any postmodernist in the community. The water in the river will bear absolutely no relation to the water consumed with dinner, or flowing from their taps, or the water meandering through Tucson in centuries gone by.

Part of the clamor of the current postmodern era is the result of competing voices. But political debate is not peculiar to the 1990s, so why should the recent political debates in the river valley be deemed postmodern? While the archaic vision and the modernists may duke it out politically, the postmodernists stay home, ignore the fray, or express total cynicism toward the outcome. For example, what were the turnout levels in some of the recent water-issue elections in Tucson? Only 37 percent of eligible voters turned out in 1977 for the recall election that ousted three city council members.[34] In other words, almost two-thirds of the voters stayed home. In 1995, when the WCPA placed restrictions on the city's use of CAP water, only 27 percent of the eligible voters bothered to cast ballots.[35] And keep in mind, this was also an election that chose the mayor and three council members, not just the so-called "hot-button" issue of CAP water. In 1999, when voters in Tucson defeated a proposition that would have strengthened the WCPA, the turnout was higher, 41 percent, but less than city officials had expected. This election also included mayoral and council seats, with the mayoral candidates especially running on both sides of the CAP issue. Given the heightened emotions at large in the community, officials had expected at least 53 percent to turn out, barely a majority of those eligible to vote.[36]

In part this is the postmodern phenomenon as I describe it in this narrative. If politics determine the fate of the river and its aquifer, and if the river and its aquifer are central to the survival of human society in the valley, then how is it that so many valley residents, at least in Tucson, can be so disassociated from the political process? I am making no claim in this environmental history to explain why so many fail to participate in electoral water politics; it is merely my assertion that the fact of low voter turnouts is one example of clear disassociation. On some level, a non-voting majority of Tucsonans considers the water politics surrounding the river and its aquifer to be irrelevant to their lives.

On an intellectual basis, postmodernism is very fond of critiquing the modernists' faith in rationalism. In this sense, much of the "irrelevance" judgment directed toward the water politics may stem from a heightened awareness of flawed logic within the policy alternatives. Take for example the regulatory efforts to control groundwater pumping in the Tucson area. The "Third Management Plan" for the Tucson AMA contains "water budget scenarios" describing the projected overdraft of groundwater in the Tucson AMA from 1995 to 2025. Without its heightened enforcement efforts, the report's "base scenario" projected an overdraft of 3,346,900 acre-feet. The "Third Plan scenario" of increasing enforcement projected an overdraft over the same period of 3,126,650 acre-feet.[37] In other words, the "tightening of the screws" would result in a reduction of the overdraft by 220,250 acre-feet, while the continued legal pumping remained over 3 million acre-feet in arrears. Such pronouncements cannot inspire confidence in the expert and scientific management of Tucson's dwindling groundwater supplies. Although it is questionable how many of those non-voters in Tucson are aware of these statistics and assertions, the end result is the same. What happens to the river and its aquifer is beyond their control, and may be beyond anyone's control. Water may be central to their lives, but water politics is irrelevant to their daily concerns—such is the post-modern vision.

Building a true and long-lasting political consensus behind any of the initiatives concerning riparian reconstruction, the CAP, or aquifer management, will be difficult. But no matter what the political outcomes, the river will remain central to valley residents. Without the river and its aquifer there would be no human society to wrangle over water politics in the Santa Cruz Valley.

Epilogue
A River Now and Then

EVEN IN ITS DRIEST REACHES, the river flows now and then, and travelers
along its course occasionally find perennial flows, although in this matter of
perennial flows it is helpful if the observer accepts treated effluent as "the
river." But not much remains of the ancient river and this concluding
survey of its current circumstances will note only a few tenuous con-
tinuities between the river's nature at the dawn of human presence or
historic record, and the river's condition now at the beginning of the
twenty-first century—a now and then river as it existed then and now.

When the Santa Cruz joins the Gila River on the Gila River Indian Reservation southwest of Phoenix, nothing is left of the Santa Cruz except thick
stands of tamarisk, desert broom, and creosote amidst countless twisting,
shallow scratches on the desert floor. In theory more than fact, the Santa
Cruz River joins the Gila River just to the northeast of what had been the
tiny community of Santa Cruz, wiped out by floods in 1993. Only in the
most severe stages of flooding does the river traverse some portion of the
jornada through the lower basin. Usually the Santa Cruz River is dry at
the confluence, and for most of the year, especially during the dry summer
months before the seasonal monsoon arrives, the Gila River is also dry.[1]

Ten thousand years ago during the moist late Pleistocene, the two rivers
no doubt joined in a wet, gushing rush, and the first humans who entered
the valley probably witnessed this opulent confluence. By historic times,
however, about 300 years ago, the region had dried significantly and the
confluence had diminished in extravagance. The Gila River still flowed,
but the Santa Cruz through the lower basin had sunk beneath the sand, and
the accounts of the dry jornada give ample evidence of the parched circumstances. But the travelers' reports also noted the seasonal variations of
the river through the lower basin, dry as a bone in May and June before the
summer rains provided water pockets along the road and commenced the

springs at the base of Picacho Peak to flowing again. Shallow wells along the road at the stagecoach stations gave the first hint of the groundwater source awaiting extraction.

Travelers today approximate the path of that traditional jornada while zooming along I-10 from Phoenix to Tucson. The interstate's designation is east-west, but from Phoenix the highway's course is more north-south, and travels nearby the Santa Cruz River. None of the ambience of the mule-borne or walking journey remains, especially as air-conditioned sedans traverse the lower basin in about an hour. But not too long ago, say in the 1950s, when cars were more likely to suffer from the desert heat, bits of the old adventure remained. Baby boomers might recall a childhood road trip, windows down to catch a 100-degree breeze, water bag tied to the door handle, watching the engine temperature rise in the old Chevy six-banger, hoping Dad wouldn't turn on the heater so as to cool off the engine (the engine's condition more important than the passengers' comfort). This was in no way the equivalent of walking from Phoenix to Tucson, but in the imagination of a youngster perhaps it captured a bit of the old jornada.

Now the interstate hums with commerce and the ribbon of development includes outlet malls, hotels, restaurants, gas stations, and easy access to retirement communities and golf courses. Agriculture seems to be receding, a few operations visible from the highway, but nothing like the postwar boom days of boundless agribusiness. In this case, appearances deceive. Take the Central Arizona Irrigation District, for example. The district is headquartered in Eloy, Arizona, and is composed of 87,000 acres of farmland. In any given season, about 60,000 acres will be cultivated, with more than 20,000 acres fallow, and most of this land is not visible from the highway. In fact only about one mile of the district's farmland fronts the highway, and bits of it are being sold to developers. Recently developers bought twenty acres for a new truck stop, and another twenty-five acres were sold for a new subdivision. Considering the overall holdings, however, the corridor of development along the interstate constitutes an insignificant diminution of the district's agricultural lands. More critical of course is the water supply. But in this area, the irrigation district remains in pretty good shape, with contracts for CAP water well into the twenty-first century and groundwater still the less-expensive alternative. Legislation regulates the amount of groundwater pumping and problems of subsidence occur. But large-scale agriculture continues apace.[2]

The CAP canal also enters the Santa Cruz Valley from north to south, not exactly in line with the old river channel, but nearby. Both the highway and the canal pass Picacho Peak to the east, while the old river channel passes the peak on the west. The highway and canal clearly intrude on the terrain in modern and postmodern ways. The highway curves and obeys general dictates of the terrain in the Santa Cruz Flats, not that far removed from the original grading and leveling of the railroad. As with the railroad construction engineers, the highway engineers sought to shave off obstructing hillsides and bridge the interrupting ravines, and the overall effect, although no less artificial, was softer than the CAP canal. The canal largely ignores terrain and defies any sense of geological order: straight lines and abrupt angles, water pumped uphill so as to flow downhill, pumped higher uphill to flow downhill again, in the aggregate a concrete-lined, staircased river flowing uphill into the middle basin.

Passing Brawley Wash, which originates in Mexico, and the farming community of Marana, the river and highway enter the middle basin through the gap between the Tucson Mountains in the west and the Tortolita Mountains in the east. The canal bends to the right and enters the Tucson area to the west of the Tucson Mountains. By the time it reaches its terminus near Tucson, CAP electric pumps have lifted the Colorado River water 2,400 feet.

At first it seems that the Santa Cruz River in the middle basin resembles something of its natural, historic form. Before reaching Camino del Oro on the northern fringe of Tucson, the channel exists in relatively non-engineered freedom, and there are modest riparian areas downstream from the city's water treatment plants. But soon the changes in the river channel become apparent. Entering the urbanized area, the cemented and stabilized river channel appears. Rillito Creek joins the river in the northern region of the basin, also cemented and stabilized, and as with the Gila River to the north, it is primarily a dry confluence. The river enters and leaves the Tucson Basin underground. In years, decades and centuries gone by, the river traditionally returned to the surface near Martinez Hill and the Black Mountains in the vicinity of the mission at San Xavier del Bac. Several springs fed into the channel, and the perennial flow meandered through a broad, shallow floodplain. Now the channel is deeply incised through the process of arroyoization, and the river only flows during seasonal floods. To protect developed property near the river, engineers and surface-water hydrologists have created a soil-cemented, straightened, and otherwise stabilized river channel through the urbanized area.

The Tucson Basin has witnessed the greatest changes in the nature and

On Tucson's northern fringe, the Santa Cruz River's perennial flow has been restored by discharge from sewage treatment plants. The flow has given rise to a modest riparian area, with the ubiquitous industrial operation occupying the floodplain.

configuration of the Santa Cruz River. Lately, the evolution of the river includes aggradation of the channel, as seasonal runoff patterns add sediment to the floor of the arroyo through Tucson. Most likely, the next big flood event will scour the channel and remove the added sediment on the arroyo floor. But even if no such scouring took place, given the degree of urbanization in close proximity to the channel, it is certain the city govern-

The Ina Road bridge spans the river channel in Tucson's northern reaches. In this recent photograph, the river seems safely contained even as a modest flood spills over the embankment on the right. During the record flood of 1983, all the bridges over the river in the Tucson Basin were closed, and for a time the river came to dominate urban society.

ment would never allow the river to return to its shallow, meandering pattern of the past, even if "natural" forces were pushing in that direction. Human engineering, no stranger to the Tucson Basin since prehistoric times, now dominates the basin to an unparalleled degree. Among the most obvious engineering features are the sculpted and reinforced riverbanks, immune, hopefully, to erosion. Less obvious, but no less significant for the river's configuration, are the hundreds of Tucson Water Department well sites scattered throughout the basin.

Decades of steady pumping have altered even the underground flow in the aquifer. For millennia the underground flow in the valley had mirrored the aboveground flow, the aquifer tending downhill toward the Gila River Basin to the north. In the 1950s the flow in the aquifer had in places shifted as underground water sought to replenish the sinking aquifer near the busiest wells. Aboveground the Santa Cruz River follows the natural contours toward the northwest. Below ground the water follows contrary gradients in all directions. The more water Tucson pumps out, the greater the rush down the underground slopes.

Entering Tucson from the north on the interstate, the river channel is often visible to the west, on the right. Passing the downtown high-rises to the east on the left, the river channel bears little resemblance to the meandering stream of the not-so-distant past, lost amidst hundreds of acres of cultivated fields. Even the more recent arroyo looks benign compared to its first appearance in the 1890s, when valuable agricultural land was being swallowed up by the migrating headcut. Now the land along the river near downtown, especially on the west bank, is largely vacant, and the subject of ambitious development plans.

Continuing on the interstate just south of the downtown area, I-10 veers to the east toward Benson along the approximate path followed by the Mormon Battalion in December 1846. As I-10 turns east, I-19 branches off to the right, heading south out of the Tucson Basin toward Nogales. It is I-19 that continues to mimic the river's path, nearby to the west. The stabilized channel extending south from downtown gives way to a generally unregulated channel beyond 36th Street, although further portions of the river are marked with engineering works and stabilization efforts. Along the stretch of the channel south of "A" Mountain, much of the river's history has been obliterated. Thankfully, many Hohokam living sites have been preserved and studied by scholars from the University of Arizona, Arizona State Museum, and Bill Doelle's Desert Archaeology, Inc., but other more recent aspects of the river's history are nowhere to be seen. Nothing remains of Silver Lake, except the name of a street. The last trace of the former main branch of the river (the existing channel occupies the former spring branch) was filled in and graded in the 1960s. By then the old channel was little more than a linear depression in the ground. Nothing remains of the old Tucson Farms Crosscut. The cultivated acres of Midvale Farms now sprout houses in a subdivision. South of 36th Street a sand and

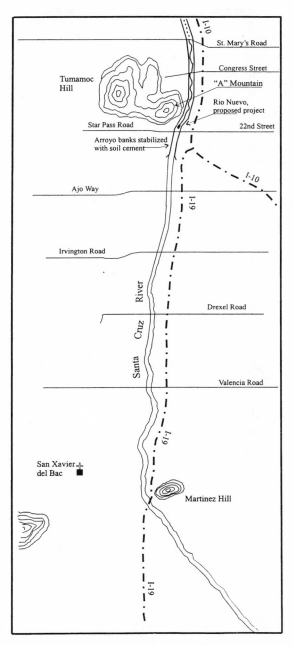

Armored with soil cement to stabilize the banks against erosion during floods, the straightened river channel today serves primarily as a flood-control mechanism, although the proposed Rio Nuevo project would return surface flow to the entrenched stream at the behest of tourism and urban renewal.

gravel plant stirs up dust in the river channel in what seems to be an appropriate industrial use for the river channel today. Dry and desolate, the river channel serves primarily as a drainage ditch for flood control. Little hint remains of the small stream's fecundity as it once meandered through Tucson.

Near San Xavier, the sweeping bend in the river is armored on its outside curve by carefully manicured bastions of stone and gravel, clearly visible to the west of the freeway. Passing nearby the slope of Martinez Hill to the east, the highway crosses the dry river channel just upstream from the stabilized curve, traversing that stretch of the river where geological formations traditionally had forced the underground flow to the surface. Now the channel is dry, and nothing remains of the shady mesquite bosque, formerly the oozing cienega, south of Martinez Hill.

The lack of surface flow near San Xavier is the result of wells and groundwater withdrawals near Sahuarita. The City of Tucson wells cannot be seen from the interstate, but the acres of pecan groves along the river channel are within clear view a mile or two to the east. The pecan groves continue south through Continental toward Canoa, but the groves and city wells are not the only draws on water along this reach of the river. The retirement community of Green Valley, an unincorporated area in Pima County, has grown in population and size, now hosting over 24,000 residents and seven golf courses. The copper mines south of Tucson have fallen into quiescence due to declining copper prices, although their tailings dumps still dominate the western horizon. But whereas the mines' use of water has declined from the booming highs of recent years, the agricultural and domestic demand has increased. Overall, these demands have completely erased any semblance of natural order to the river as it enters the middle basin from the south. Where geology traditionally dictated returning surface flows, no water runs.

Passing Green Valley, the highway soon enters Santa Cruz County, and hints of the renewed perennial flow appear to the east. The geology of bajada slopes and the river channel at their base is clearly seen as the interstate winds generally south. But also clearly seen are the smattering of houses and traces of development along the highway, the small, white block of an observatory atop Mt. Hopkins in the Santa Rita Mountains, and countless other additions to the scene that remove the view from its nineteenth-century and earlier appearance. Nonetheless, of the three basins, the river appears least changed in the upper basin. From Nogales Wash north, the effluent from the international wastewater treatment plant replenishes the river's flow, albeit in a manner that stretches the concept of a naturally flowing stream. However, the cottonwood and willow trees seem to human eyes to appreciate the steady flow, and the effluent discharge has the virtue of constancy. This constancy remains contingent on the Mexican flow into the system, a circumstance that is in no way guaranteed indefinitely. The flow is subject to regulation by the International Boundary and Water Commission. There is also talk of establishing "sharing" agreements between municipal, agricultural, and other potential users of treated effluent (seeking to establish something akin to Tucson's water-recycling efforts, which diminish the flow out of its treatment plants), but these agreements to share the discharge from the plant in the upper basin remain ideas for the future, not a policy in force today. As of now, the river flows steadily north from Nogales. In the not-so-distant past, perennial flow in the river ended at Canoa, and today the flow from the treatment plant north of Nogales almost makes it that far. Fourteen miles downstream from the wastewater treatment plant, the river passes Tubac and Tumacacori. The riparian area in this stretch of the river has taken on a sense of great age that belies its newness. It seems natural, and is defended as such by local activists. Whether natural or artificial, the trees are thick and luxurious, reminiscent of the shady groves of centuries gone by. From the elevated position of the interstate, the repaired dome of the mission at Tumacacori—now a National Historic Monument—protrudes from amidst a blanket of vegetation. Along this stretch, the river's flow truly does seem opulent.

Just south of Tumacacori the highway makes a sweeping curve up and then drops down to the floodplain along the river. This is the place where the 1983 flood washed out the road, and in the reconstruction, engineers placed the new highway high above any future erosion. The interstate continues south toward Nogales on the floodplain, the bajada slopes now above the roadbed. Nearing the Sonoita Creek confluence, the development of Rio Rico comes into view, occupying the bluffs overlooking the creek and river. The Rio Rico community of several thousand homes (2,166 homes in 1994, with hopes to increase the community by 200 homes per year thereafter)[3] occupies mostly slopes on the north side of Sonoita Creek on the east side of the river, whereas the railroad town of Calabasas occupied the southern bluffs. The golf course is mostly in the floodplain of Sonoita Creek. Additions to the flow in the river from Sonoita Creek drainage, once sizeable, are now mostly nonexistent. The canyon is plugged by the dam forming Lake Patagonia, and even though the dam releases at least fifty acre-feet a month as a "base flow," no surface water reaches the river. The settlement and resort of Rio Rico routinely pump thousands of acre-feet from the aquifer, and so any surface flow in the creek quickly sinks beneath the sand. Only during flood episodes does the river gain visibly from the tributary drainages of the San Cayetano Mountains in the east and Tumacacori and Atascosa Mountains in the west.

From the Nogales Wash, the river turns southeast past the site of the old Spanish mission at Guevavi, and the interstate continues south toward the international border. A dirt road on the southern bank follows the river during its eastward traverse, passing the old mission site on the opposite bank of the river. Here the stream sometimes rises in the channel briefly as bedrock ascends once again, but more times than not the shallow aquifer is so depleted that the channel remains dry along this stretch, upstream and cutoff from the effluent flow emanating from Nogales Wash. The aquifer is depleted because the Nogales, Arizona, wells are just upstream, and the demands of the growing city place a consistent draw on the groundwater beneath the channel. Continuing east along the graded county road—River Road, of course—often shaded by the trees that still line the river-banks, the old Nogales pumping plant appears, constructed in 1912 and still housing the original infiltration galleries, on the southern bank of the river at the mouth of Proto Canyon. Here the dirt road ends where Arizona State Highway 82, heading north to Patagonia, crosses the river.

The crossroads at Proto Canyon and the river is marked by a little bustle of development, a few houses on the north bank of the river, the "little red school house" (still mostly red, but not so little anymore), Nogales International Airport on the bluffs to the north overlooking the river, and the community of Kino Springs along the river and its floodplain upstream from the pumping plant. A paved road into the development to the east follows along the river channel. Within a mile the ubiquitous golf course appears, occupying a portion of the floodplain, with the clubhouse, restaurant, and new garden homes trailing generally upstream.

A few miles beyond the last houses in Kino Springs, the river crosses the border into Mexico, flowing once again as barely a trickle as bedrock rises to shallow depths. Cottonwoods line the banks. In its thirty-two-mile passage through Mexico, the river provides the domestic water needs of Nogales, Sonora, from the Paredes and Mascarenas wellfields, with the addition of water from the Magdalena watershed. These domestic uses are significant, but the modest flow at the border persists in spite of the wells. If proposals by Mexican officials to increase pumping at the well fields by 45 percent come to pass, the cottonwood trees at the border would soon disappear. The river also serves the struggling communal farming community of San Lázaro, but agriculture in the floodplain still comes under varying federal regulation, depending on the political winds emanating from Mexico City. The riparian area persists in areas along this stretch of the river, and residents of San Lázaro work to preserve what is left of the flowing stream. Their efforts will succeed only if agriculture remains at the current modest levels.[4]

The village of Santa Cruz remains isolated at the end of a long, bumpy dirt road, and the river, so recently within a stone's throw of an interstate highway, now courses largely unobserved. The river is used by the farmers and ranchers of Santa Cruz, with only a hint of the rapaciousness of their American colleagues downstream. The river near Santa Cruz appears as a broad and shallow creek in the service of livestock and agriculture. As an indication of the draw on the river, the cienega that gave Bartlett so much trouble outside of the town is gone, and the river in places goes dry, but through most of its reach through Santa Cruz the small, trickling stream remains alive in the broad, sandy channel. Beyond Santa Cruz, nearing the international border, the little brook quickly becomes more river-like, thirty feet wide and five feet deep. This impressive stream diminishes before the border crossing, however. With the rising elevations at the southern end of the San Rafael Valley, the flow in the river slackens, and the stream crosses the border as barely a rivulet two miles east of Lochiel.[5]

Least changed in appearance over the ten millennia or so of human presence in the valley is the San Rafael Valley in the upper basin. The hills forming the river's watershed have escaped the most severe woodcutting and the grassland in the valley the most severe overgrazing during the periodic crashes in the cattle industry. The river is merely a stream in the San Rafael Valley, but the terrain remains remarkably pristine compared to other vistas along the river's course.

The greatest threat to the river in the San Rafael Valley appears to be the prospect of ranching's demise. Breaking up the large ranches into smaller parcels would result in more wells, more development, and ultimately, more people, at least, such was the concern of long-time rancher Bob Sharp: "The system is set up to break up private land. That's the bottom line, and that's what is going to destroy the river . . . [with] more wells, more pumping of ground water, we could dry up the riparian [area] if we had too much development." Sharp, and other valley ranchers, tried to work legalistic and political maneuvers to their own, and the river's, advantage. The pattern of development foreseen by Sharp has already taken place in the area of Nogales. As recently as forty years ago, grazing cattle welcomed the river's return to U.S. territory. In the 1960s, Kino Springs and its golf course supplanted ranching activity, and current efforts to foster development around the Nogales, Arizona, airport seem to indicate the further loss of pastoral landscapes.[6]

Moving up the San Rafael Valley from south to north, the low Canelo Hills rise as blue swells in the distance. The river still begins in those hills, little more than a seep and ooze at first, growing quickly into a modest stream. Although the start of the river's journey remains largely unchanged, and the stretch through the high valley greatly resembles the scene from centuries gone by, most of the river's course is a testament to change. Human society has wrought many of those changes, but natural forces of climate and hydrology have influenced the river's circumstances as well. The river has never stood still, even as it has disappeared through most of its reaches downstream. The river will continue to flow, aboveground or below, through periods of ebb and flow, as the forces of change act upon it.

I sometimes wonder if my great grandparents would recognize the river that drew them to the valley so long ago. At milepost twelve, where their house once stood, the scene along the river is not that much changed. Cottonwood trees still line a shallow channel with a perennial stream. Of course the stream is perennial only because of a consistent flow of treated effluent, and the hills overlooking the scene are speckled with the homes of Rio Rico. "Rich River," or lessening stream, the assessment has always been in the eye of the beholder.

Notes

Introduction

1. Smith, *Groundwater Supply and Irrigation*, 118.

2. Webb, *The Great Plains*.

3. Parsons defines *bioregion* as "a geographical province of marked ecological and often cultural unity, its subdivisions, at least ideally, often delimited by watersheds of major streams." Parsons, "On 'Bioregionalism' and 'Watershed Consciousness,' " 1.

4. For an introduction to bioregional studies, see Parsons, "On 'Bioregionalism' and 'Watershed Consciousness,' " 1–6; Flores, "Place: An Argument for Bioregional History," 1–18; Frenkel, "Environmental Determinism and Bioregionalism," 289–295; McTaggart, "Bioregionalism and Regional Geography," 307–319; Alexander, "Bioregionalism: Science or Sensibility?" 161–173. Also, "Bioregionalism and Bioregional History," a panel of three papers on Pacific Northwest history presented at the American Society for Environmental History Conference, Tacoma, Washington, March 15–19, 2000.

5. Worster, *Rivers of Empire*. Worster directs readers to the work of Karl Wittfogel—for example, "Hydraulic Approach to Pre-Spanish Mesoamerica." See also Karl Butzer, *Early Hydraulic Civilization in Egypt* and *Environment and Archeology*.

6. Parker, "Channel Change," 1. Charles Bowden cites 8,990 square miles in "Death of the Santa Cruz Calls for a Celebration." Julio Betancourt describes the "total basin area" as 23,300 square kilometers (13,980 square miles) in "Arroyo Legacy," 4.

Chapter 1. The River as It Was

1. Stokes and Judson, *Introduction to Geology*, 316–320, 335–337; Tellman, Yarde, and Wallace, *Arizona's Changing Rivers*, 7.

2. In southern Arizona, the mountains include crystalline rocks from the Precambrian period as well as "volcanic, plutonic, and metamorphic rocks that range from Mesozoic to Cenozoic age, [and] some sedimentary rocks that range from Paleozoic to Cenozoic age." Parker, "Channel Change," 5–6; Stokes and Judson, *Introduction to Geology*, 257.

3. Field et al., "Geomorphic," 36; Parker, "Gravity Analysis," 53; Helmick, "Santa Cruz River Terraces," 10–12.

4. Field et al., "Geomorphic Survey," 36.

5. Czaplicki and Mayberry, *Archaeological Assessment,* 7.

6. "There is evidence that faulting has at least continued into the late Pleistocene and even into the present. However, in southeastern Arizona it appears that the main phase of the disturbance ended within the interval from latest Miocene to early Pliocene (6 to 3 m.y.b.p.), and generally quiet tectonic conditions have persisted into the present." Helmick, "Santa Cruz River Terraces," 11.

7. Fenneman, *Physiography of Western United States,* 379–380.

8. Coggeshall, "Hydrologic Assessment," 25.

9. Parker, "Channel Change," 6.

10. Helmick, "Santa Cruz River Terraces," 10.

11. Parker, "Gravity Analysis," 53.

12. "Late Pliocene and Pleistocene," in Field et al., "Geomorphic Survey," 36; "Middle Pleistocene," in Betancourt, "Tucson's Santa Cruz River and the Arroyo Legacy," 33.

13. Parker, "Channel Change," 6.

14. "Evidence for a large main channel flowing parallel to the present Santa Cruz is lacking. Exposures of the basin fill are relatively high in the piedmont and have been partly eroded during burial. It is possible that the river may have been smaller or even flowed southward into Mexico rather than in its present valley." Helmick, "Santa Cruz River Terraces," 13.

15. Betancourt, "Arroyo Legacy," 33.

16. Coggeshall, "Hydrologic Assessment," 25–26; Davis, "Canelo Hills Cienega," 35.

17. Parker, "Gravity Analysis," 53.

18. Gelt et al., *Water in the Tucson Area,* 25.

19. Coggeshall, "Hydrologic Assessment," 38.

20. Gelt et al., *Water in the Tucson Area,* 17.

21. Betancourt, "Arroyo Legacy," 24–26; see also Webb and Betancourt, *Climatic Variability and Flood Frequency,* 10–27. Cooke and Reeves also studied arroyo formation, focusing on human-caused factors as part of "several interrelated environmental changes." Cooke and Reeves, *Arroyos and Environmental Change,* 47–55, 97–99.

22. Czaplicki and Mayberry, *Archaeological Assessment,* 8–9; Helmick, "Santa Cruz River Terraces Near Tubac," 7.

23. As Paul Martin theorized, the Paleo-Indian hunters may have hunted scores of species of large mammals to extinction, participating in what he called "Pleistocene Overkill." Martin, "Pleistocene Overkill," 32–38; Martin and Klein, eds., *Quaternary Extinctions.*

24. Sheridan, *Arizona: A History,* 5–6.

25. Gelt et al., *Water in the Tucson Area,* 1.

26. A mammoth killed by eight spear points of the Clovis culture is the oldest

evidence of human presence in the region, in the San Pedro Valley to the east. The mammoth bones show no sign of butchering, however, perhaps indicating that it escaped from its tormentors even though mortally wounded. Sheridan, *Arizona: A History*, 3.

27. Martin, "Pleistocene Overkill," 32.

Chapter 2. From Clovis Points to Maize

1. The first evidence of human presence in the Santa Cruz River valley is of the Clovis complex, around 9500 B.C. Bryan and MacNeish assume a human presence in the Southwest prior to the Clovis culture. Bryan, *Paleo-American Prehistory*; MacNeish, "Early Man in the New World"; Czaplicki and Mayberry, *Archaeological Assessment*, 15, 23; Gregonis and Huckell, *The Tucson Urban Study*, 20; Fontana, "Calabazas of the Rio Rico," 66.

2. The only Folsom and Plano evidence in Arizona occurs in the far northeastern corner of the state. Czaplicki and Mayberry, *Archaeological Assessment*, 15−16.

3. Ibid., 19, 23, 25. Although most deposits occurred near cienegas, Hewitt and Stephen reported a site in the Tortolita Mountains in an "oak-juniper-mixed cactus zone around 1,219 meters in elevation." Hewitt and Stephen, *Archaeological Investigations*.

4. Doelle, "Human Use of the Santa Cruz River in Prehistory"; Czaplicki and Mayberry, *Archaeological Assessment*, 42.

5. DiPeso, Haury, and Gregonis and Reinhard suggest a "continuum model," which postulates little population movement and an evolutionary link between prehistoric Hohokam and historic Piman groups. DiPeso, *San Cayetano del Tumacácori*, 434; Haury, *The Hohokam: Desert Farmers and Craftsmen*, 5, 265, 321, 351−357; Gregonis and Reinhard, *Hohokam Indians of the Tucson Basin*, 4. Ezell agrees with the model somewhat tentatively. Ezell, "Is There a Hohokam Continuum?" 61−66. Czaplicki and Mayberry caution: "At present much more research is necessary before the origins of the Historic Native American groups of southern Arizona can be established, including their relationship, if any, to the prehistoric Hohokam." Czaplicki and Mayberry, *Archaeological Assessment*, 60. As William H. Doelle explained, "The relationship of the folks who were farming along the Santa Cruz at 1000 B.C. is not absolutely clear. They are probably the ancestors of the later Hohokam, but that is not fully established to everyone's satisfaction." Doelle, "Human Use."

6. For example, "When Kino first visited San Xavier and Tucson in the 1690s, the Indian population in the Santa Cruz Valley was greater than any other area in southern and central Arizona, although still much reduced from the Hohokam maximum." Betancourt, "Arroyo Legacy," 36.

7. Haury, *The Hohokam: Desert Farmers and Craftsmen*, 266; Kelly, *The Hodge's Ruin*, 78−82.

8. Gelt et al., *Water in the Tucson Basin,* 3.

9. Burkham, *Depletion of Streamflow,* 5; Czaplicki and Mayberry, *Archaeological Assessment,* 7; Fenneman, *Physiography of Western United States,* 379–380.

10. Fish and Fish, "Hohokam Political and Social Organization," 157.

11. For a comparative study of societies and their irrigation systems, see Downing and Gibson, eds., *Irrigation's Impact on Society.* Of particular interest in Hohokam studies is "Human Use," by William H. Doelle, the most prominent scholar on the Hohokam in the Tucson area. See also the essay by Wayne Kappel, "Irrigation Development and Population Pressure," 159–167; Meyer, *Water in the Hispanic Southwest,* 12; Fish and Nabhan, "Desert as Context," 49; Ackerly, Howard, and McGuire, "La Ciudad Canals," 9–11; Nials, Gregory, and Graybill, "Salt River Streamflow," 70–74; Downum, "Between Desert and River," 17. Evidence for Hohokam canals also exists in the Upper Basin in the vicinity of the Sedgwick Ranch. *Arizona Daily Star,* October 4, 1994, A6.

12. Meyer, *Water in the Hispanic Southwest,* 12, 15.

13. Fish and Nabhan, "Desert as Context," 47–48. See also Bowden, *Killing the Hidden Waters.*

14. For example, "The interval A.D. 950–1150 is one of the best documented periods of increased effective moisture on the Colorado Plateaus." Euler et al., "The Colorado Plateaus," 1096; Drew, *Tree-Ring Chronologies of Western America,* 27–28; Dean and Robinson, *Dendroclimatic Variability,* 81–86. McGuire and Schiffer advise caution in using Colorado Plateau data: "Although large-scale drought episodes probably covered most of the Southwest, extensive variability in the intensity and range of such episodes has been documented. These studies indicate that severe droughts occurring on the Colorado Plateau may not have occurred in the Sonoran Desert and vice versa." McGuire and Schiffer, eds., *Hohokam and Patayan Prehistory,* 51. The yearly variation in rainfall by as much as 55 percent is a factor in this circumstance as well. Gelt et al., *Water in the Tucson Area,* 3.

15. Czaplicki and Mayberry, *Archaeological Assessment,* 39.

16. Downum, "Between Desert and River," 16; Fish et al., "Prehistoric Agave Cultivation in Southern Arizona," 100, 107–112; Coggeshall, "Hydrologic Assessment," 23; Greenleaf, *Excavations at Punta de Agua,* 16–17; Arnold, "Vertebrate Animals," 28–43, 49–53; Rothrock, *Preliminary Botanical Report,* 126; Thornber, "Plant Acclimatization in Southern Arizona," 1–9; Czaplicki and Mayberry, *Archaeological Assessment,* 8.

17. U.S. Surgeon General's Office, *A Report on Barracks and Hospitals,* 462–463; William H. Doelle, e-mail communication with the author. September 8, 1999.

18. Betancourt, "Arroyo Legacy," 73–74.

19. Jamie Whelan has found that native populations in the Southeast utilized a much wider variety of wildlife species for their sustenance than Europeans recognized. Whelan, "From Archives and Trash Pits." This is corroborated in the Southwest from the chronicles of early Spanish travelers, who reported native populations

eating "delicacies" such as "snakes, lizards, locusts, worms, caterpillars, and all kinds of insects." Pfefferkorn, *Sonora: Descriptions of the Province,* 190.

20. Officer, *Hispanic Arizona,* 195–200.

21. Pfefferkorn, *Sonora: Descriptions of the Province,* 195–197.

22. At one site, Emil Haury used living sites to estimate a population of 2,000 people plus or minus 50 percent. Later scholars lowered the population estimate to 1,000, then 600, and finally 500 individuals. Haury, *The Hohokam: Desert Farmers and Craftsmen,* 75–77, 356; Wilcox, McGuire, and Sternberg, *Snaketown Revisited,* 185–197; Plog, "Explaining Culture," 14.

23. Fish and Fish, "Hohokam Political and Social Organization," 157; Haury, *The Hohokam: Desert Farmers and Craftsmen,* 150. Haury's drawings indicate ten or eleven rebuildings.

24. The water consumption of cattle varies according to how much they eat, their salt intake, and ambient temperature. In general, however, during hot summer months cattle require about ten gallons a day, while human beings require about one gallon a day. Adolph, *Physiology of Man in the Desert,* 112; Hicks et al., *Water Intake by Feedlot Steers,* 209.

25. McKee and Wolf, eds., *Water Quality Criteria,* 112; Fish and Nabhan, "Desert as Context," 51; Downum, "Between Desert and River," 17; Adolph, *Physiology of Man in the Desert,* 112; Blaney and Harris, *Consumptive Use and Irrigation,* 32; Coggeshall, "Hydraulic Assessment," 22; Meyer, *Water in the Hispanic Southwest,* 78.

26. Fish and Fish, "Hohokam Political and Social Organization," 165.

27. Ibid.

28. Crown and Judge, *Chaco and Hohokam,* 263, 303.

29. Tellman, Yarde, and Wallace, *Arizona's Changing Rivers,* 11.

30. Doelle, Huntington, and Wallace, "Rincon Phase Reorganization," and Gregory, "The Morphology of Platform Mounds," 71–95, 184; Fish and Fish, "Hohokam Political and Social Organization," 165–166.

31. These counts are derived from maps in William H. Doelle and Henry D. Wallace, "Hohokam Regional System," 332–333. Historic and prehistoric Indians also occupied several villages in the upper basin. These sites generally are not considered in studies of the Hohokam due to the lack of archaeological evidence. The presumed sites are buried under several meters of alluvium.

32. Doelle, "Human Use."

33. See U.S. Census reports, and Logan, *Fighting Sprawl and City Hall.*

34. Nials, Graybill, and Gregory, "Salt River Streamflow," 70–74; Doelle and Wallace, "Hohokam Regional System," 333; Betancourt, "Arroyo Legacy," 36; Schwalen and Shaw, *Ground Water Supplies.*

Chapter 3. Cattle, Wheat, and Peace

1. I have chosen to follow James Officer's spelling of Spanish names through the Spanish and Mexican periods. With the incursion of increasing numbers of Anglos

in the 1850s, Spanish place names became somewhat Anglicized. For example, the accent was dropped from Tumacácori, and the "z" changed for an "s" in Calabazas. The reader should also note that scholars at times differ in their use of accent marks and other symbols. For example, Officer, Meyer and others spell Guevavi without further notation, while Campa and others give the name as Güévavi. Campa, *Hispanic Culture, 59.*

2. Officer, *Hispanic Arizona,* 31. Father Kino assigned the first priests to the Pima settlements of Guevavi and Bac in 1701. In 1732, another round of assignments placed priests at Bac, Guevavi and at Santa María Soamca (later known as Santa Cruz). Kessell, *Mission of Sorrows,* 28–29, 43. Hammon provides a translation of the letter reporting the 1732 assignments in Hammond, "Pimería Alta After Kino's Time," 227–235. Doelle and Wallace, "Hohokam Regional System," 334.

3. Juan Mateo Manje's last name is sometimes spelled Mange. See, for example, Campa, *Hispanic Culture, 61.*

4. Bolton, *The Padre on Horseback,* 36; Smith, Kessell, and Fox, *Father Kino in Arizona,* 75; Polzer, *Kino Guide II,* 19.

5. Burrus, *Kino and Manje,* 215–219. Manje's transcribed reports from the fourth expedition, November 2 to December 6, 1697, are on pages 333–384 of Burrus et al., *Father Kino in Arizona,* 43–44. In 1692, Kino had reported "eight hundred souls" at San Xavier. Oblasser, "Papaguería: The Domain of the Papagos," 6. Czaplicki and Mayberry, *Archaeological Assessment,* 60–61.

6. Fontana described the archaeological sites in Fontana, "Calabazas of the Rio Rico," 66. Dobyns described the "ranchería" living pattern in Dobyns, *Tubac through Four Centuries.* Robinson agrees that most natives in the upper basin did not live in a "riverine, village settlement pattern" but rather evidenced a "Papago-like ranchería adaptation." Robinson, "Mission Guevavi: Excavations in the Convento," 172. See also Kessell, *Mission of Sorrows, 51.*

7. Smith, Kessell, and Fox, *Father Kino in Arizona,* 49, 57.

8. The processes of "ecological imperialism" as described by Alfred Crosby took place in the valley. For example, Dobyns extensively studied measles and smallpox epidemics in Sonora; see Dobyns, *Spanish Colonial Tucson,* 32, 105, 139; *Tubac through Four Centuries,* 16–18, 32, 40–45, 99–106, 139. See also Dobyns, "Indian Extinction," 163–181; Meyer, *Water in the Hispanic Southwest,* 78–79.

9. MacCameron, "Environmental Change in Colonial New Mexico"; de la Teja, *San Antonio de Bexar.*

10. Gelt et al., *Water in the Tucson Area,* 6; MacCameron, "Environmental Change in Colonial New Mexico," 82.

11. White, "Ecological Crisis," 341–351.

12. Officer cited Kessell, who in turn cited Bolton's brief reference: "After the destruction of the village of Motolicachi some ten years previously [1687], several stock ranches near the Pima border had been abandoned. . . . One of these was San Lazaro, the property of Romo de Vivar, of Bacanuche." Bolton, *Rim of Christendom,*

358n. 2. Officer mentioned that Romo de Vivar "also had property in Cananea and at the south end of the Huachuca Mountains as early as 1680." Officer, *Hispanic Arizona,* 31–32; Kessell, *Mission of Sorrows,* 51n. 12. Kessell quoted an improbable reference by Juan Mateo Manje to the level of ranching in northern Sonora in the 1690s: "Already these enemies [the Apaches] had devastated and consumed the ranches (*estancias*) of Terrenate, Batepito, Janos, and San Bernardino, where there had been more than one hundred thousand head of cattle and horses." Kessell, "The Puzzling Presidio," 22.

13. Officer, *Hispanic Arizona,* 31–32; Hastings, "People of Reason and Others"; Meyer, *Water in the Hispanic Southwest,* 55.

14. Smith, Kessell, and Fox, *Father Kino in Arizona,* 13–14; Meyer, *Water in the Hispanic Southwest,* 55, 61; Oblasser, "Papaquería," 3–9; Kessell, *Mission of Sorrows,* 92.

15. Officer cites a fanega as equal to 3.2 bushels. Lt. Cave Johnson Couts described a fanega as 2.5 bushels in 1848. Officer, *Hispanic Arizona,* 119; Couts, *Hepah, California,* 54–55.

16. Bannon, *The Spanish Borderlands Frontier,* 70; Oblasser, "Papaguería," 6; Officer, *Hispanic Arizona,* 33–34; Meyer, *Water in the Hispanic Southwest,* 31, 41. Meyer described the acequias as "ditches [which] ranged in depth from two to eight or nine feet and in width from a foot to seven feet." For comparison, the Hohokam canals in the Phoenix Basin were 180 miles long, up to thirty feet wide and ten feet deep.

17. Pfefferkorn, *Sonora,* 197; Officer, *Hispanic Arizona,* 41, 66, 147–148. The prevalence of mescal was reported in Safford and Hughes, "The Story of Marianna Diaz."

18. Meyer, *Water in the Hispanic Southwest,* 78.

19. Pfefferkorn, *Sonora,* 289–290.

20. de la Teja, *San Antonio,* 8–9; Worcester, "The Apache Indians of New Mexico in the Seventeenth Century," 1, 19; Wissler, *Indians of the United States,* 224–228; White, *"It's Your Misfortune,"* 10, 12–14.

21. In 1768 the Indians in Tucson were "mostly Sobaipuri refugees from the San Pedro Valley." Kessell, *Friars, Soldiers and Reformers,* 56; Officer, *Hispanic Arizona,* 48, 73; McCarty, *A Spanish Frontier,* 66.

22. Residents of Tubac, after the presidio moved to Tucson, "reported that the Apaches were openly grazing stolen horses in the Santa Cruz Valley." Officer, *Hispanic Arizona,* 54; Parke, *Report on Explorations for Railroad Routes,* appendix C.

23. The policy decision to move the Sobaipuri was made in 1761 and carried out in 1762. The confusion over the numbers of Sobaipuri involved in the transfer is described by Dobyns in his translation of the report on the move made by Francisco Elías Gonzáles in March 1762: "Their number reaches 250, although the missionary and justice informed me that they numbered 400 souls." Dobyns, *Spanish Colonial Tucson,* 19–21; Kessell, *Mission of Sorrows,* 160–162. In the late 1760s, Indian raids

by the Apaches and Comanches also caused the abandonment of ranches in Texas near San Antonio. de la Teja, *San Antonio,* 100.

24. Meyer, *Water in the Hispanic Southwest,* 56, 61.

25. Kessell, *Friars, Soldiers and Reformers,* 39; Officer, *Hispanic Arizona,* 47; Meyer, *Water in the Hispanic Southwest,* 64.

26. Kessell cited a 1774 census that said Guevavi and Sonoita were abandoned by that time. Kessell, *Friars, Soldiers and Reformers,* 88; *Mission of Sorrows,* 119, n.50. Robinson said the mission was abandoned in 1773. Robinson, "Mission Guevavi," 135, 171; Baldonado, "Missions of San Jose de Tumacácori," 21–24; Meyer, *Water in the Hispanic Southwest,* 62, 90.

27. Baldonado, "Missions San Jose de Tumacácori," 23.

28. Officer, *Hispanic Arizona,* 51.

29. Without the garrison in the presidio at Tubac, the Hispanic population gradually abandoned the area. A census in 1783 "based on earlier figures" reported that Tubac was abandoned: "its 158 persons having left for the Colorado River settlements or the presidio of Tucson because of water shortage and Apaches." At Tumacácori 125 residents remained in forty-two Indian and seven Spanish families. Calabazas had eighty-four residents in forty families. Kessell, *Friars, Soldiers and Reformers,* 154n. 18; Officer, *Hispanic Arizona,* 58.

30. Meyer, *Water in the Hispanic Southwest,* 56; Dobyns, *Spanish Colonial Tucson,* 67.

31. Dobyns, *Spanish Colonial Tucson,* 73.

32. In New Mexico, MacCameron found some linear decline in environmental conditions from the early Colonial period but also found cyclical patterns of change as well as site-specific circumstances that determined environmental conditions. MacCameron, "Environmental Change in Colonial New Mexico," 96–97; Hicks et al., *Water Intake by Feedlot Steers,* 209.

33. Dobyns, *Spanish Colonial Tucson,* 79, 86–87; Officer, *Hispanic Arizona,* 63, 66.

34. Kessell, *Friars, Soldiers and Reformers,* 164; Officer, *Hispanic Arizona,* 63, 66.

35. Meyer stated, "I have been unable to locate a copy of the agreement, but reference to it is made in at least two subsequent documents, one dated 1796 and one dated 1821." Meyer, *Water in the Hispanic Southwest,* 56n. 31. The 1796 reference appears in Friar Bringas's report. Matson and Fontana, *Friar Bringas Reports to the King,* 66.

36. Kessell placed the Papagos at El Pueblito. Kessell, *Friars, Soldiers and Reformers,* 197. Meyer reported that the Papagos settled at San Xavier del Bac. Meyer, *Water in the Hispanic Southwest,* 61.

37. Matson and Fontana, *Friar Bringas Reports to the King,* 66, 72; Meyer, *Water in the Hispanic Southwest,* 56–57, 62, 90; Kessell, *Friars, Soldiers and Reformers,* 185–186, 197–198.

38. The low prices for livestock limited the ability of the priest at Tumacácori, Father Narciso Gutiérrez, to finance the building of a new church. The church built

by the Jesuits in 1757 was a "crumbling ruin" by the early 1800s. Officer, *Hispanic Arizona,* 79. Kessell described the priest's dilemma: "They could try to raise surplus wheat, but that depended on the weather. The mission did have livestock, more than ever before. But prices had fallen off sharply. Cattle that sold just five years earlier for ten pesos a head, now brought only three and a half . . . [and] the price might soon drop to a peso." Kessell, *Friars, Soldiers and Reformers,* 202.

39. The weaving trade at Tumacácori prospered until an Apache raid in 1801 decimated their sheep herd. Kessell, *Friars, Soldiers and Reformers,* 202. McCarty, *Desert Documentary,* 89.

40. Tucson's commander valued the presidio's corn crop slightly higher than de León had priced Tubac's crop: 1,800 pesos for 600 bushels in Tucson, 1,200 pesos for 600 bushels in Tubac.

41. McCarty, *Desert Documentary,* 82–90.

42. Indians were under no illusions about the effect grazing cattle had on limited water supplies. A hundred years earlier, Kino had faced resistance from the Indians at Remedios in central Sonora because they realized the Spanish "pastured so many cattle that the watering places were drying up." Meyer, *Water in the Hispanic Southwest,* 61; Fontana, *Of Earth and Little Rain,* 45.

43. McCarty, *Desert Documentary,* 85–87, 92.

44. Officer, *Hispanic Arizona,* 87–88.

45. Ibid., 93, 352n. 50.

46. Kessell, *Friars, Soldiers and Reformers,* 247; Officer, *Hispanic Arizona,* 88.

47. Kessell pointed out that earlier "folksy" histories were mistaken in their claim that the Indians' grant was much larger, "over 52,000 acres." Kessell, *Friars, Soldiers and Reformers,* 213, 214n. 77. Wagoner described the Indians' grant as "about 5,000 acres." Wagoner, *Early Arizona,* 219; Arizona (Territory) Surveyor General, *Journal of Private Land Grants.*

48. Wagoner, *Early Arizona,* 211; Kessell, *Friars, Soldiers and Reformers,* 291; Arizona (Territory) Surveyor General, *Journal of Private Land Grants.*

49. Kessell acknowledged the increasing population, with qualifications: "Gente de Razón entered the valley in increasing numbers during the first two decades of the nineteenth century, not at a constant rate but when conditions dictated." Kessell, *Friars, Soldiers and Reformers,* 238. Officer noted the relatively rare deaths by the Apaches in 1819. Officer, *Hispanic Arizona,* 87, 351n. 26.

50. Dobyns, *Spanish Colonial Tucson,* 102–104.

51. Ibid., 139; Officer, *Hispanic Arizona,* 87.

52. Meyer, *Water in the Hispanic Southwest,* 145–164; Officer, *Hispanic Arizona,* 93.

Chapter 4. War and a Returned River

1. Information on all the grants in the Santa Cruz Valley can be found in Mattison, "Early Spanish and Mexican Settlements in Arizona," 285–327. See also

Willey, "La Canoa: A Spanish Land Grant Lost and Found," 154; Meyer, *Water in the Hispanic Southwest,* 81; and Hadley and Sheridan, *Land Use History of the San Rafael Valley,* 22–24.

2. Wagoner, *Early Arizona,* 159–161, 229.

3. Kessell, *Friars, Soldiers and Reformers,* 245, 279–281; Officer, *Hispanic Arizona,* 106–108.

4. Officer, *Hispanic Arizona,* 109–110, 149, 171.

5. McCarty, *A Frontier Documentary,* 34, 68–70.

6. Ibid., 34.

7. Officer, *Hispanic Arizona,* 151, 154.

8. Weber, *The Taos Trappers,* 112.

9. Weber said the Indians reported to "Commandante Pacheco." Officer said the report reached "Tucson Commanding Officer Manuel de Leon," and that Mayor Ygnacio Pacheco informed the Sonoran governor of the Americans' presence. Ibid., 121; Officer, *Hispanic Arizona,* 105.

10. McCarty translated the report on the Americans' visit to Tucson made by Juan Romero, Tucson's "third constitutional mayor." McCarty pointed out that Romero had replaced Ignacio Pacheco as mayor on January 1, 1827, the day after the Americans entered Tucson. So the Americans may have reported to Pacheco, but it was Romero who informed the Sonoran government of their presence. McCarty, *A Frontier Documentary,* 8; Marshall, "St. Vrain's Expedition," 151–160.

11. Betancourt suggests that the reference to trapping on the Santa Cruz may have been to muskrats, an animal similar to the beaver except that muskrats do not construct dams. Betancourt, "Arroyo Legacy," 54, 72–73.

12. Weber, *The Taos Trappers,* 123–124.

13. Meyer, *Water in the Hispanic Southwest,* 143; Officer, *Hispanic Arizona,* 111; McCarty, *A Frontier Documentary,* 16–17.

14. Meyer, *Water in the Hispanic Southwest,* 57; Officer, *Hispanic Arizona,* 113. McCarty translated Escalante's report in *A Frontier Documentary,* 13–15.

15. Officer, *Hispanic Arizona,* 114; McCarty, *A Frontier Documentary,* 16–18.

16. Officer, *Hispanic Arizona,* 114–115, 262–263.

17. Ibid., 111–114; Kessell, *Friars, Soldiers and Reformers,* 244, 279–281.

18. Officer, *Hispanic Arizona,* 120.

19. Kessell, *Friars, Soldiers and Reformers,* 280–281.

20. McCarty translated the 1831 census for Tucson, Tubac, and Santa Cruz in *Copper State Bulletin,* 93.

21. Potash translated the Ramirez report in "Notes and Documents," 332–335.

22. Officer, *Hispanic Arizona,* 110, 123, 125.

23. Ibid., 127, 361n. 36; Kessell, *Friars, Soldiers and Reformers,* 290–291.

24. Hadley and Sheridan, *Land Use History,* 24; Officer, *Hispanic Arizona,* 148–149, 171.

25. Officer, *Hispanic Arizona*, 148.

26. Ibid., 133, 166, 363n. 60; McCarty, *A Frontier Documentary*, 14.

27. Officer, *Hispanic Arizona*, 159, 161.

28. Ibid., 166–167; Kessell, *Friars, Soldiers and Reformers*, 302.

29. McCarty, *A Frontier Documentary*, 90.

30. Ibid., 90–91; Officer, *Hispanic Arizona*, 166–169, 370n. 45.

31. Officer, *Hispanic Arizona*, 185, 191.

32. Ibid., 120; McCarty, *A Frontier Documentary*, 90–91.

33. McCarty, *A Frontier Documentary*, 41.

34. Officer, *Hispanic Arizona*, 167.

35. Kessell, *Mission of Sorrows*, 29.

36. Meyer, *Water in the Hispanic Southwest*, 92.

37. McCarty, *A Frontier Documentary*, 12–15.

38. Ibid., 90.

39. Ibid.

Chapter 5. Americanos

1. Bigler, "Extracts from the Journal," 46–47; Bliss, "Journal," 79–80; Cooke, *Conquest*, 175; Whitworth, "Mississippi to the Pacific," 127–160.

2. Bliss "Journal," 80; Cooke, *Conquest*, 145; Bigler, "Extracts from the Journal," 48.

3. Bliss, "Journal," 80–81; Bigler, Bigler, "Extracts from the Journal," 48–49; Officer, *Hispanic Arizona*, 197–199.

4. Bigler, "Extracts from the Journal," 49.

5. Ibid; Cooke, *The Conquest*, 150; Golder, *March of the Mormon Battalion*, 195–196.

6. McCarty, *A Frontier Documentary*, 119; Officer, *Hispanic Arizona*, 206, 208–209.

7. Kessell, *Friars, Soldiers and Reformers*, 246; Officer, *Hispanic Arizona*, 215.

8. Officer, *Hispanic Arizona*, 209, 215.

9. Ibid., 210–211; Dobyns, "Prologue," in Couts, *Hepah, California*, 2–3.

10. Couts, *Hepah, California*, 54.

11. Ibid., 54–55.

12. Ibid., 59.

13. Ibid., 60. Dobyns points out that Graham's troops no doubt had raided the farms of the Papagos, who tilled unfenced plots in the floodplain. Ibid., 60n. 22.

14. Ibid., 61–62; Chamberlain, *My Confession*, 257–258.

15. Couts, *Hepah, California*, 62. Chamberlain's description of the weather was quite different. He talks about being handcuffed in the sun for a minor transgression: "This Tucson is a very hot place, and being tied up in the sun I was in a fair way of becoming *tasajo* (jerked beef). I could feel my brains 'sizzling' and my skin was commencing to crack." This seems unlikely in late October—warm perhaps, but

not "sizzling." Officer warns that Chamberlain's account "is a confusing jumble of times and places" and readers should be wary. Chamberlain, *My Confession*, 258.

16. Couts, *Hepah, California*, 63.

17. Ibid; Officer, *Hispanic Arizona*, 380n. 28.

18. Kessell, *Friars, Soldiers and Reformers*, 308; Officer, *Hispanic Arizona*, 217, 381n. 36.

19. Officer, *Hispanic Arizona*, 218.

20. Ibid., 220–221.

21. Nevins quotes Frémont's journal in Nevins, *Frémont: The West's Greatest Adventurer*, vol. 2, 418.

22. Wood, "Personal Recollections," 9. The edited version of the journal by Goodman appeared in 1955. Officer, *Hispanic Arizona*, 382n. 12.

23. The group of Mississippians is mentioned in A. B. Clarke's journal. Clarke's group was a few days behind the New Orleans and Mississippi parties at Santa Cruz, but caught up to the other groups as they approached the Gila River. Clarke, *Travels in Mexico and California*, 87.

24. Officer, *Hispanic Arizona*, 224; Durivage, *Daily Picayune*, August 7, 1849, 1.

25. Ibid.

26. Ibid.

27. Ibid.

28. Powell, *Santa Fe Trail to California*, 145–147; Clarke, *Travels in Mexico and California*, 81–82; Pancoast, *A Quaker Forty-Niner*, 233.

29. Clarke, *Travels in Mexico and California*, 81–86; Cox, "From Texas to California in 1849," 142–143; Powell, *Santa Fe Trail to California*, 145–147; Evans, *Mexican Gold Trail*, 147–151.

30. Durivage, *Daily Picayune*, August 7, 1849, 1; Cox, "From Texas to California in 1849," 144; Pancoast, *A Quaker Forty-Niner*, 233; Powell, *Santa Fe Trail to California*, 136.

31. Clarke, *Travels in Mexico and California*, 81–82; Cox, "From Texas to California in 1849," 143; Eccleston, *Overland to California*, 199–204; Officer, *Hispanic Arizona*, 229–230.

32. Beiber, *Southern Trails to California*, 319–320; Pancoast, *A Quaker Forty-Niner*, 233.

33. Evans, *Mexican Gold Trail*, 147–151.

34. Cox, "From Texas to California in 1849," 142–143; Beiber, *Southern Trails to California*, 319–320.

35. Cox, "From Texas to California in 1849," 142–143.

36. Officer, *Hispanic Arizona*, 224.

Chapter 6. New Borders

1. Officer, *Hispanic Arizona*, 246, 249–250, 385n. 4.

2. Ibid., 253.

3. Ibid., 253–254, 262–263; Arizona (Territory) Surveyor General, *Journal of Private Land Grants;* Sheridan, *Los Tucsonenses,* 27–28.

4. Officer, *Hispanic Arizona,* 249–250.

5. Ibid., 254.

6. Ibid.

7. Ibid., 255; Bartlett gives an account of the Mormon experience in Tubac in Bartlett, *Personal Narrative,* 119–120, 304–305.

8. An effort to raise sheep for a commercial market was new to the Santa Cruz River valley but long established in the upper Rio Grande Valley of New Mexico. MacCameron, "Environmental Change in Colonial New Mexico," 83–84.

9. Fontana, "Calabazas of the Rio Rico," 77–79. A sketch of the Gandara hacienda in 1854 at "Calabaza—Santa Cruz Valley, Sonora," appears in Bailey, ed., *The A. B. Gray Report,* 82; Froebel, *Seven Years Travel,* 495–496.

10. Officer, *Hispanic Arizona,* 275, 279; Emory, *Report,* 118–119.

11. For a brief history of the tumultuous political history of the boundary surveys, see Goetzmann's introduction to Emory, *Report,* ix–xxx.

12. Bartlett, *Personal Narrative,* 256.

13. Ibid., 398–401.

14. Graham, *Report of Lt. Col. Graham,* 44.

15. Ibid., 407–408.

16. Graham, *Report of Lt. Col. Graham,* 46–47.

17. Bartlett, *Personal Narrative,* vol. 2, 252–260, 286–287.

18. Ibid., 293.

19. Ibid., 295–296.

20. Ibid., 297.

21. Ibid., 301.

22. Ibid., 302–304.

23. Ibid., 118–120, 304–305.

24. Ibid., 304, 307.

25. Ibid., 307–308.

26. Ibid., 310.

27. Cumming, "A Rancher Looks at His Environment," 72.

28. Bartlett, *Personal Narrative,* 311.

29. Ibid., 317, 320–322.

30. Parke, *Report on Explorations,* 7–8.

31. Ibid.

32. Ibid.

33. Bailey, *The A. B. Gray Report,* 75–78.

34. Ibid., 80, 84.

35. Bell, "Texas-California Cattle Trail," 308–309.

36. Ibid., 311; Officer, *Hispanic Arizona,* 393n. 53.

37. Bell, "A Log of the Texas-California Cattle Trail," 314–316.

38. Emory, *Report,* 94.

39. Ibid., 118.

40. Ibid.

41. Ibid., 118–119.

42. Officer, *Hispanic Arizona,* 283, 286–288; Sheridan, *Los Tucsonenses,* 30–31.

Chapter 7. New Residents, Old Problems

1. Pederson, "A Yankee in Arizona," 129–130; Sonnichsen, *Tucson,* 47–48; Betancourt, "Arroyo Legacy," 52. Betancourt describes the origins and descriptions of the water sources feeding Silver Lake as "muddled." References to the Rowletts appear primarily in relation to William S. Grant, who bought the lake and mill from the Rowletts in 1860. As Pederson explains, the Rowletts appear in the 1860 census for Tucson and then disappear from the record. Other historians and scholars, such as C. L. Sonnichsen, cite Maish and Driscoll as the originators of the lake. This error stems from reliance on the reminisces of C. C. Wheeler, who wrote pioneer tales for the *Arizona Daily Star* in the 1930s, and credited Maish and Driscoll, the proprietors of a resort hotel at the lake in the 1880s, with its original construction. Wheeler, "History and Facts Concerning Warner and Silver Lakes," 853. Willey's article on the La Canoa land grant also contributed to the confusion. Willey refers to Maish's hotel in town, the Palace, as being opened in 1875, with the operation of the "roadhouse and family resort at Silver Lake" as commencing "later." Willey, "La Canoa," 162.

2. Biographical material on Sam Hughes can be found in Samuel Hughes Collection, MS 366, Arizona Historical Society, particularly box 1; Jose de Castillo Collection, MS 140, Arizona Historical Society; *Tucson Citizen,* April 15–17, 20, 1915.

3. Smith, *Groundwater Supply and Irrigation,* 98; Martin, *The Lamp in the Desert,* 68.

4. Pederson, "A Yankee in Arizona," 129–141.

5. "Cultivated Fields," Fergusson 1862.

6. Fergusson, *Letter of the Secretary of War,* 14. Betancourt also misidentified Fergusson's "200 yards" as "200 m." Betancourt, "Tucson's Santa Cruz River and the Arroyo Legacy," 56.

7. Poston, *Building a State in Apache Land,* 71–72.

8. Ibid., 72–73.

9. Duffen, "Overland via 'Jackass Mail,'" 280.

10. Poston, *Building a State in Apache Land,* 73–74, 80.

11. Duffen, "Overland via 'Jackass Mail,'" 36.

12. Ibid., 157–158.

13. Ibid., 159–160.

14. Ibid., 161–162.

15. Ibid., 159; *Tucson Magazine,* December 1948. The 1948 reference says 1870s,

but Way's reference to well water was in 1858. Gustafsen, ed., *John Spring's Arizona*, 128; Gelt et al., *Water in the Tucson Area*, 6–7.

16. Duffen, "Overland via 'Jackass Mail,'" 360–361, 369.

17. Sheridan, *Arizona*, 67.

18. Ibid.; Poston, *Building a State in Apache Land*, 92, 95–96, 101–105.

19. The reader should note Browne's spelling of Tumacacori, dropping the accent mark. As more Anglos entered the region in the ensuing decades, such modifications of Hispanic place names became common. Browne, *Adventures in Apache Country*, 152–154.

20. Ibid., 191–193; Gustafson, *John Spring's Arizona*, 128–137.

21. Hadley and Sheridan, *Land Use*, 49.

22. Ibid., 64; *Tombstone Epitaph*, December 16, 1882, quoted in Murbarger, *Ghosts of the Adobe Walls*, 135.

23. Hadley and Sheridan, *Land Use*, 97–98; Wayland, "Experiment on the Santa Cruz," 44.

24. Altshuler, "The Regulars in Arizona in 1866," 119.

25. Gustafson, *John Spring's Arizona*, 47; *Arizona Daily Star*, February 28, 1891; *Weekly Arizonan*, May 22, 1869.

26. U.S. Surgeon General's Office, *A Report on Barracks and Hospitals*, 462–463.

27. Bell, *New Tracks in North America*, 335.

28. Parke, *Report on Explorations*, 7–8; Betancourt, "Arroyo Legacy," 74.

29. Foreman, Maps and Notes, 1871.

30. *Arizona Citizen*, August 22, 1874, October 30, 1875.

31. Ibid., May 2, 1879.

32. Ibid.; Barter, *Directory of City of Tucson for the Year 1881*; Warner, papers, AHS; Sonnichsen, *Tucson*, 84.

33. *Tucson Magazine*, December 1948.

34. Peterson, "Fort Lowell, A.T.," 10.

35. *Arizona Weekly Star*, August 9, 1877; Smith, *Groundwater Supply and Irrigation*, 102–103.

36. *Arizona Weekly Star*, July 18, 1878; Smith, *Groundwater Supply and Irrigation*, 98.

37. Sheridan, *Los Tucsonenses*, 68–69; Smith, *Groundwater Supply and Irrigation*, 98. Betancourt describes as "uncertain" the degree to which depletion of ground cover contributed to the flood. Betancourt, "Tucson's Santa Cruz River and the Arroyo Legacy," 86.

Chapter 8. Steel Rails and Steam Pumps

1. Hadley and Sheridan, *Land Use*, 98.

2. Manriquez, "A Brief History of the Mascarenas Ranch," 87–88.

3. Hadley and Sheridan, *Land Use*, 100.

4. Brewster, "The San Rafael Cattle Company," 138, 140–141, 149.

5. Hadley and Sheridan, *Land Use,* 101.

6. Ibid., 101, 103; Wayland, "Experiment on the Santa Cruz," 28–29, 73.

7. Hadley and Sheridan, *Land Use,* 101. Miller used the Swamp and Overflow Act by Congress to acquire thousands of acres of land during seasonal flooding in California's central valley in the 1860s. See Reisner, *Cadillac Desert,* 46.

8. Hadley and Sheridan, *Land Use,* 101.

9. Ibid., 103.

10. Brewster, "The San Rafael Cattle Company," 140, 151.

11. Hadley and Sheridan, *Land Use,* 105–106; Cooke and Reeves, *Arroyos and Environmental Change,* 47–55, 97–99; see also Bahre, *Legacy of Change;* Betancourt, "Arroyo Legacy"; Hastings, Rodney, and Turner, *The Changing Mile,* 43–46.

12. Myrick, *Railroads of Arizona,* 280. From the incursion of Anglo residents in the valley in the 1850s, the spelling of Calabazas became variegated. By the 1880s, the railroad town of Calabasas cemented the shift from "z" to "s."

13. Ingram, Laney, and Gillilan, *Divided Waters,* 30–31; Rodriguez, "The Otro Lado—Nogales, Sonora," 4.

14. Browell, "The Presidio of Tubac," 11.

15. Seibold, "Patagonia, Then and Now," 15–16; Myrick, *Railroads of Arizona,* 264; Brooks, "River Sustains," A11.

16. *Arizona Daily Star,* August 9, 1890, 3.

17. Warner Papers, AHS, 2.

18. Betancourt, "Arroyo Legacy," 91, 162.

19. A transcript of the trial is in the papers of Charles R. Drake, AHS, 39.

20. Drake Papers, AHS, 6, 7, 13–14, 19, 39.

21. Betancourt, "Arroyo Legacy," 96 (cites *Arizona Daily Star,* June 11, 1885, not in UA microfilm); Sonnichsen, *Tucson,* 111–112.

22. Pisani, *Water, Land, and Law in the West,* 17; Meyer, *Water in the Hispanic Southwest,* 148–150; Worster, *Rivers of Empire,* 89–92.

23. Willey, "La Canoa," 164; Smith, *Groundwater Supply and Irrigation,* 99–100.

24. *Arizona Daily Citizen,* September 3, 1887, 1; Betancourt, "Arroyo Legacy," 102.

25. *Arizona Daily Star,* August 13, 14, 1890, 4.

26. Ibid., August, 7, 1890, 4.

27. Ibid., August 9, 1890. 4.

28. Ibid., August 26, 1890, 4.

29. Ibid., August 9, 1890, 4.

30. Spalding, "Desert Plants," 9.

31. Olberg and Schanck, *Irrigation and Flood Protection* 7–8; Smith, *Groundwater Supply and Irrigation,* 99.

32. Tellman, Yarde, and Wallace, *Arizona's Changing Rivers,* 21; Parker, "Channel Change," 7; Betancourt, "Arroyo Legacy," 102–103, 108–109.

33. An unlikely reference to a steam pump and well at an Oro Valley ranch in

1874 appears in Roscoe Willson's *Pioneer Cattlemen of Arizona*. Willson's brief biographies of area ranchers relied on the reminiscences of the pioneers' children and grandchildren. Specific date references are sometimes problematic. To posit a steam pump almost immediately upon acquisition of the ranch and livestock by the owner, George Pusch, and six years prior to the arrival of the railroad is not impossible, but it is unlikely, given the difficulty and expense of transporting the mechanism overland by horse- and mule-drawn freight. However, the 1874 date has been repeated in Kupel, "Diversity through Adversity," 43; and Gregonis and Huckell, *The Tucson Urban Study*, 46–47, both citations relying on the Willson citation.

34. *Arizona Daily Star*, February 5, 1891, 4.

35. Peterson, "Fort Lowell, A.T.," 10; Gelt et al., *Water in the Tucson Area*, 6.

36. *Arizona Daily Star*, January, 16, 1892, 4; May 1, 1902, 5.

37. Allison, "Pioneer Days in Tucson," 16.

38. Ibid., 16–17; *Arizona Daily Star*, May 1, 1902, 5; Kupel, "Diversity through Adversity," 88–90.

39. *Arizona Daily Star*, January 16, 1892, 4.

40. Betancourt, "Arroyo Legacy," 121.

41. *Arizona Daily Star*, September 24, 1882, 4; Gelt et al., *Water in the Tucson Area*, 7.

42. Warner Papers, AHS.

43. *Arizona Daily Star*, October 24, 1891, 4.

44. Betancourt, "Arroyo Legacy," 103.

45. *Arizona Daily Star*, July 16, 1891, 4.

46. Gelt et al., *Water in the Tucson Area*, 7.

47. Most references cite 1893 as the year of the water company's first well, although Gelt et al., *Water in the Tucson Area*, gives 1889 as the year. This seems unlikely given the above-average rainfall in Tucson that year—over eighteen inches—and a search of newspapers and water company records shows no reference to a well in 1889.

48. *Arizona Daily Star*, March 3, 21, 31, 1893, p. 4 in each issue; *Charter and Ordinances of the City of Tucson*, 59.

49. *Arizona Daily Star*, April 27, 1893, 4; June 11, 1893.

50. Ibid., July 15, 1887, 4; *Arizona Daily Citizen*, September 12, 1887, 4.

51. Olberg and Schanck, "Irrigation and Flood Control," 10.

52. *Arizona Daily Star*, July 13, 1887, 4.

53. Ibid., December 20, 1892, 4. Olberg and Schanck described the Indians' wood supply as 6,440 acres of "timber land," that could be turned to agriculture, given enough water for irrigation. Olberg and Schanck, "Irrigation and Flood Control," 5.

54. Smith, *Groundwater Supply and Irrigation*, 203; Kupel, "Diversity through Adversity," 87.

55. A miner's inch is "a unit of measure of water flow, varying with locality but

often a flow equaling 1.5 cu. ft. per minute." *Random House Dictionary,* 2d edition, 1224.

56. *Arizona Daily Star,* February 12, 1884, 4.

57. Olberg and Schanck noted that 1,580 acres "were said to have been under irrigation," in 1890, while Castetter and Bell cited 720 acres under cultivation. Olberg and Schanck, *Irrigation and Flood Protection* 5; Castetter and Bell, *Pima and Papago Indian Agriculture,* 50–51.

58. Olberg and Schanck, *Irrigation and Flood Protection* 5.

Chapter 9. Water Mining and the Wagging Urban Tail

1. Reisner, *Cadillac Desert,* 157–158.

2. Davis, "Las Vegas versus Nature," 53–73; Hundley, *The Great Thirst;* Reisner, *Cadillac Desert.*

3. Sonnichsen, *Colonel Greene,* 232–239.

4. Hadley and Sheridan, *Land Use,* 129

5. Ibid., 130.

6. *Arizona Cattlelog,* December 1953, 32–33.

7. Hadley and Sheridan, *Land Use,* 112–114, 139–140, 142.

8. Ibid., 139–140, 274. Hadley and Sheridan cited the Parker quote from the Parker family manuscript, "in [the] possession of several Parker family members."

9. Ibid., 142.

10. Schwalen and Shaw, *Ground Water Supplies,* 77; Browell, "The Presidio of Tubac," 12; Cumming, "A Rancher Looks at His Environment," 74, 76; Brooks, "Sonora Leg Nourishes Rural Lives," A7.

11. Schwalen and Shaw, *Ground Water Supplies,* 98–100.

12. Ingram, Laney, and Gillilan, *Divided Waters,* 60; Myrick, *Railroads of Arizona,* 296; Horton interview.

13. Ingram, Laney, and Gillilan, *Divided Waters,* 55, 60.

14. Ibid.

15. Allison, "Pioneer Days in Tucson," 17.

16. *Arizona Daily Star,* May 1, 1902, 5.

17. Smith, *Groundwater Supply and Irrigation,* 201–202.

18. *Arizona Daily Star,* July 10, 1902, 1. Another trumpeting of the potential water supply appeared in the newspaper, July 30, 1902, 4.

19. James, *Arizona: The Wonderland,* 317.

20. Ibid., 312; Betancourt, "Arroyo Legacy," 130, 155. Betancourt cited 6,000 acres by 1913, while James claimed the company bought 12,000 at its inception in 1910.

21. Schwalen and Shaw, *Ground Water Supplies,* 94.

22. Hinderlider, "Irrigation of Santa Cruz Valley," 200–201, 242; James, *Arizona: The Wonderland,* 313.

23. Jones interview.

24. Hinderlider, "Irrigation of Santa Cruz Valley," 244

25. Schwalen and Shaw, *Ground Water Supplies,* 75, 82–83; Kupel, "Diversity through Adversity," 106.

26. Olberg and Schanck, *Irrigation and Flood Protection* 11–12.

27. *Arizona Daily Star,* June 17, 1923, 1; July 13, 1923, 3; May 3, 1949, 2.

28. Ibid., February 23, 1940, 3; *Beat the Peak,* Tucson Water Sources; Davidson, *Geohydrology and Water Resources.*

29. *Charter and Ordinances,* 288–289; *Arizona Daily Star,* February 13, 1923, 3; Gelt et al., *Water in the Tucson Area,* 8.

30. *Arizona Daily Star,* February 13, 1923, 3; June 8, 1923, 2. Potential reservoir sites had been surveyed in the area since 1899. In that year investors studied Bear Canyon, and in 1901 they announced the first plan for a dam in Sabino Canyon. The dam was to be 200 feet high at a site in the canyon where the walls were only thirty feet apart. Another dam site at Tanque Verde was surveyed in 1906. Smith, "Groundwater Supply and Irrigation, 122–124.

31. Smith, *Groundwater Supply and Irrigation,* 118, 195.

32. *Arizona Daily Star,* June 19, 1903, 5.

33. Ibid., February 15, 1893, 2.

34. Ibid., June 13, 1893, 4; June 23, 1893, 4.

35. Miller, *The Urbanization of Modern America,* 52–53.

36. *Arizona Daily Star,* September 7, 1921, 3; September 21, 1921, 2; February 6, 1923, 3.

37. Ibid., February 6, 1923, 3.

38. Ibid., March 13, 1923, 2; May 22, 1923, 2; May 23, 1923, 1; May 24, 1923, 8.

39. Smith, *Groundwater Supply and Irrigation,* 125.

40. Ibid.

41. Betancourt, "Arroyo Legacy," 167.

42. *Arizona Daily Star,* June 24, 1905, 2; Olberg and Schanck, *Irrigation and Flood Protection* 8. The same siphoning phenomenon and draining of the water table occurred in the Rillito channel. Smith, *Groundwater Supply and Irrigation,* 97–100.

43. Betancourt, "Arroyo Legacy," 146–158.

44. Ibid., 155.

45. Ingram, Laney, and Gillilan, *Divided Waters,* 60.

46. Betancourt, "Arroyo Legacy," 169–170.

46. Ibid., 143–146.

47. Ibid., 166–171.

48. Sonnichsen, *Tucson,* 220.

Chapter 10. Lost Balance and "The Truth About Water in Tucson"

1. Schwalen and Shaw, *Ground Water Supplies,* 77; Browell, "The Presidio of Tubac," 12; Cumming, "A Rancher Looks at His Environment," 74; Brooks, "Sonora Leg Nourishes Rural Lives," A7.

2. Hadley and Sheridan, *Land Use,* 142.

3. *Arizona Cattlelog,* December 1953, 32–33.

4. Cumming, "A Rancher Looks at His Environment," 76.

5. *Arizona Cattlelog,* December 1953, 35–36.

6. Manriquez, "Mascarenas Ranch," 88.

7. Ingram, Laney, and Gillilan, *Divided Waters,* 62.

8. Browell, "The Presidio of Tubac," 12–13.

9. Ingram, Laney, and Gillilan, *Divided Waters,* 62.

10. Ibid., 95.

11. Tellman, Yarde, and Wallace, *Arizona's Changing Rivers,* 23.

12. Cushman, "Lower Santa Cruz Area," 115–135.

13. Gelt et al., *Water in the Tucson Area,* 20–21.

14. Cushman, "Lower Santa Cruz Area," 77.

15. Ibid., 81; Schwalen and Shaw "Ground Water," 77.

16. Betancourt, "Arroyo Legacy," 166–171.

17. *Arizona Daily Star,* February 23, 1940, 3.

18. Lazaroff, *Sabino Canyon,* 98–99; Logan, *Fighting Sprawl and City Hall,* 20.

19. Sonnichsen, *Tucson,* 258; *Tucson Citizen,* October 30, 31, 1936.

20. Gelt et al., *Water in the Tucson Area,* 8.

21. Betancourt, "Arroyo Legacy," 170–171; Aldridge and Eychaner, "Floods of October 1977," 15–18. Tellman and Gelt maintain that the mesquite bosque survived into the 1960s. Tellman, Yarde, and Wallace, *Arizona's Changing Rivers,* 23; Gelt et al., *Water in the Tucson Area,* 9.

22. Brandt, *Arizona and Its Bird Life,* 71–72.

23. Gelt et al., *Water in the Tucson Area,* 8–9.

24. Davidson, *Geohydrology and Water Resources,* 37.

25. Betancourt, "Arroyo Legacy," 170–171.

26. *Arizona Daily Star,* February 23, 1940, 2, 3, 6, 7, 10, 13; October 1, 1940, 1.

27. Ibid., December 8, 1941, 1, 3.

28. Sonnichsen, *Tucson,* 272.

29. Halpenny, *Gila Basin,* 7–8.

30. Ibid.

31. Johnson, "Upper Santa Cruz Basin," 105–113.

32. "The Truth About Water in Tucson," UASC.

33. Schwalen and Shaw, *Ground Water Supplies,* quote in the Foreword without page number.

34. Ibid., 2.

35. Ibid., 89, 25–70, 71.

36. Gelt et al., *Water in the Tucson Area,* 9.

37. *Arizona Daily Star,* November 6, 1959, B1.

38. *Arizona Daily Star,* February 9, 1954, 7; February 10, 1954, 3; May 21, 1957, B1; October 24, 1962, B1.

39. Ibid., March 2, 1949, A8; March 22, 1949, A2; July 19, 1949, 3; August 21, 1949, A1–2; August 22, 1949, A3; August 23, 1949, A1, 6; August 24, 1949, A1.

40. Ibid., April 6, 1949, A1.

41. Ibid., February 3, 1925, 3; August 5, 1952, A3; August 6, 1959, A1; August 25, 1959, A2; April 5, 1960, B1; April 13, 1960, A8; April 21, 1960, B8; Gelt et al., *Water in the Tucson Area,* 9.

42. *Arizona Daily Star,* May 17, 1949, 2; April 21, 1955, 10.

43. Betancourt, "Arroyo Legacy," 180–181.

Chapter 11. Preservation and Conservation

1. Patagonia-Sonoita Creek Preserve, n.p.

2. Arnold, "Vertebrate Animals," 28–43; Rothrock, *Preliminary Botanical Report,* 126; Thornber, "Plant Acclimatization in Southern Arizona," 1–9.

3. Manriquez, "Mascarenas Ranch," 88.

4. Ibid., 89–90.

5. Rodriguez, "The Otro Lado—Nogales, Sonora," 4–5; Horton interview.

6. Aldridge and Eychaner, "Floods of October 1977," 74; Cumming, "A Rancher Looks at His Environment," 74.

7. Cumming, "A Rancher Looks at His Environment," 74.

8. Census of Agriculture, 1950, 1954; Cushman, "Lower Santa Cruz Area," 115–135.

9. Census of Agriculture, 1950, 1959, 1982.

10. Ibid., 1954, 1959, 1964, 1974, 1978, 1982.

11. Tellman, Yarde, and Wallace, *Arizona's Changing Rivers,* 21; Gelt et al., "Water in the Tucson Basin," 9.

12. Gelt et al., "Water in the Tucson Basin," 9.

13. Logan, *Fighting Sprawl and City Hall,* 47.

14. Tellman, Yarde, and Wallace, *Arizona's Changing Rivers,* 19.

15. Ibid., 21.

16. Davidson, *Geohydrology,* 37, 40.

17. Ibid.

18. Ibid., 42; Halpenny also mentioned "cones of depression" in 1952. Halpenny, *Gila Basin,* 21.

19. Tellman, Yarde, and Wallace, *Arizona's Changing Rivers,* 23.

20. Davis, "Las Vegas versus Nature," 71.

21. *Arizona Daily Star,* November 4, 1976, 1.

22. Reisner, *Cadillac Desert,* 271–274; Young, *The Arizona Water Controversy;* Young and Martin, "Economics of Arizona's Water Problem."

23. *Arizona Daily Star,* November 4, 1976, 1; January 1, 1977, 1; January 16, 1977; January 19, 1977; Gelt et al., *Water in the Tucson Area,* 10.

24. Gelt et al., *Water in the Tucson Area,* 10.

25. *Beat the Peak.*

26. Ibid.

27. Ibid.

28. *Arizona Daily Star,* February 6, 1923, 3; Tellman, Yarde, and Wallace, *Arizona's Changing Rivers,* 22.

29. Aldridge and Eychaner, "Floods of October 1977," 4, 7–8, 10–13.

30. Parker, "Channel Change," 25–26.

31. *Arizona Daily Star,* October 2, 1983, A-2; October 3, 1983, A-1,2.

32. Saarinen et al., *Flood of October 1983,* 101–108. See also Roeske, Garrett, and Eychaner, "Floods of October 1983."

33. Ibid.

34. Baker, "Questions Raised," 65–66. Baker questioned the effectiveness of bank stabilization efforts in his conference paper: "From an overall river management perspective, piecemeal bank protections generates greater channel instability than does no protection at all." See also Montes, "Effects of the October 1983 Discharge," 51.

35. Saarinen et al., *Floods of October 1983,* 26.

36. Betancourt, "Arroyo Legacy," 187–189.

37. Ponce, Osmolski, and Smutzer, "Large Basin Deterministic Hydrology," 1227–1245.

38. "Flood Insurance Study," FEMA; Betancourt, "Arroyo Legacy," 182, 184.

39. Betancourt, "Arroyo Legacy," 181, 185.

40. Johnson, "Basic Data Report," 38.

41. Tellman, Yarde, and Wallace, *Arizona's Changing Rivers,* 123–124.

42. Ibid., 153.

43. Gelt et al., *Water in the Tucson Area,* 16.

44. Ibid., 10.

Chapter 12. CAP, Broken Pipes, and Groundwater Spikes

1. Johnson, *The Central Arizona Project;* Welsh, *How to Create a Water Crisis;* Gelt et al., *Water in the Tucson Area,* 11.

2. Coate, "Biggest Water Fight," 86; Gelt et al., *Water in the Tucson Area,* 11.

3. Sheridan, *Arizona,* 219–227.

4. Ibid.

5. August, *Vision in the Desert,* 91–92, 100, 201.

6. Gelt et al., *Water in the Tucson Area,* 11.

7. Ibid., 11–12.

8. Ibid., 12–13.

9. Johnson, "Basic Data Report," 1, 4–5, 38.

10. Gelt et al., *Water in the Tucson Area,* 10; University of Arizona President Peter Likins, quoted in the Foreword to *Water in the Tucson Area,* i.

11. Ibid., 20–21.

12. Ibid., 21–22.

13. Ibid., 17.

14. Pence and Brooks, "Santa Cruz: On the River," A10.

15. Lin Lawson, state water biologist is quoted in Brooks, "Human Demands Put Strain on Flow," A6.

16. Ingram, Laney, and Gillilan, *Divided Waters,* 88; Tellman, Yarde, and Wallace, *Arizona's Changing Rivers,* 22; Brooks, "Human Demands Put Strain on Flow," A6.

17. Horton interview.

18. Ibid.

19. Brooks, "Human Demands Put Strain on Flow," A6.

20. Ibid.

21. McEachern interview; Long interview.

22. Long interview.

23. Census of Population, 1950, 1960, 1970, 1980, 1990.

24. For information on the initial negotiations, see *A Profile of Arizona's San Rafael Valley,* and *A Framework for Guiding the Future of the San Rafael Valley.*

25. *Arizona Daily Star,* January 22, 1999.

26. Ibid.

27. Brooks, "River Sustains," A11; "Sonora Leg Nourishes Rural Lives," A6.

28. Horton interview.

29. *Arizona Daily Star,* January 13, 1999.

30. Ibid., April 26, 1999.

31. *Third Management Plan,* 10–6, 10–7.

32. *Arizona Daily Star,* May 28, 1999; August 13, 1999.

33. Ibid., June 22, 1999.

34. Ibid., January 19, 1977.

35. Ibid., November 8, 1995, 1A.

36. Ibid., November 3, 1999, 12A.

37. *Third Management Plan,* 10–7.

Epilogue

1. Cushman, "Lower Santa Cruz Area," 115–116; Pence, "Arid Ditch," A1.

2. McEachern interview.

3. Brooks, "Human Demands Put Strain on Flow," A6.

4. Ibid., A1; Brooks, "Sonora Leg Nourishes Rural Lives," A6.

5. Brooks, "Sonora Leg Nourishes Rural Lives," A1.

6. Pence and Brooks, "Santa Cruz: On the River," A1, A10.

Bibliography

Ackerly, Neal W., Jerry B. Howard, and Randall H. McGuire. *La Ciudad Canals: A Study of Hohokam Irrigation Systems at the Community Level.* 7 vols. Arizona State University, Anthropological Field Studies No. 17, 1985.

Adolph, E. F. *Physiology of Man in the Desert.* New York: Interscience Publishers, Inc., 1947.

Aldridge, B. N., and J. H. Eychaner. "Floods of October 1977 in Southern Arizona and March 1978 in Central Arizona." U.S.G.S. Water-Supply Paper 2223, 1984.

Alexander, Donald. "Bioregionalism: Science or Sensibility?" *Environmental Ethics* 12 (Summer 1990): 161–173.

Allison, Warren. "Pioneer Days in Tucson." Unpublished manuscript compiled by Elvira R. Odom, in Allison Family Papers, Arizona Historical Society, MS 113.

Altshuler, Constance Wynn, ed. "The Regulars in Arizona in 1866: Interviews with Henry I. Yohn." *Journal of Arizona History* 16:2 (Summer 1975): 119–126.

Arizona (Territorial) Surveyor General. *Journal of Private Land Grants.* University of Arizona Library Microfilm No. 2174.

Arizona Weekly Star.

Arnold, Lee W. "An Ecological Study of the Vertebrate Animals of the Mesquite Forest." M.S. thesis, University of Arizona, 1940.

August, Jack L. *Vision in the Desert: Carl Hayden and Hydropolitics in the American Southwest.* Fort Worth: Texas Christian University Press, 1999.

Bahre, C. J. *Legacy of Change: Historic Human Impact on Vegetation in the Arizona Borderlands.* Tucson: University of Arizona Press, 1991.

Bailey, L. R., ed. *The A. B. Gray Report: Survey of a Route on the 32nd Parallel for the Texas Western Railroad, 1854.* Los Angeles: Westernlore Press, 1963.

Baker, Victor R. "Questions Raised by the Tucson Floods of 1983." *Journal of the Arizona-Nevada Academy of Science.* Vol. 19, Annual Conference Report, April 1984.

Baldonado, Luis. "Missions of San Jose de Tumacácori and San Xavier del Bac in 1774." *The Kiva* 24:4 (April 1959): 21–24.

Bannon, John Francis. *The Spanish Borderlands Frontier, 1513–1821.* Albuquerque: University of New Mexico Press, 1974.

Barter, G. W. *Directory of the City of Tucson for the Year 1881.* San Francisco: G. W. Barter, 1881.

Bartlett, John Russell. *Personal Narrative of Explorations and Incidents in Texas, New Mexico, California, Sonora, and Chihuahua.* 2 vols. New York: D. Appleton and Company, 1854.

Beat the Peak: A Foundation for Water Conservation Education in Tucson. Tucson Water Sources, Tucson Water Company, 1996.

Beiber, Ralph P., ed. *Southern Trails to California in 1849.* Vol. 5. Glendale, Calif.: Arthur H. Clarke Company, 1937.

Bell, James G. "A Log of the Texas-California Cattle Trail, 1854." J. Evetts Haley, ed. *Southwestern Historical Quarterly* 35:4 (April 1932): 290–316.

Bell, William A. *New Tracks in North America: A Journal of Travel and Adventure Whilst Engaged in the Survey for a Southern Railroad to the Pacific Ocean during 1867–8.* London: Chapman and Hall, 1870; reprint, Albuquerque: Horn and Wallace Publishers, 1965.

Betancourt, Julio. "Tucson's Santa Cruz River and the Arroyo Legacy." Ph.D. dissertation, University of Arizona, 1990.

Bigler, Henry W. "Extracts from the Journal of Henry W. Bigler." *Utah Historical Quarterly* 5:2 (April 1932): 35–64.

Blaney, Harry F., and Karl Harris. *Consumptive Use and Irrigation Requirements of Crops in Arizona.* U.S. Department of Agriculture, Soil Conservation Service, 1951.

Bliss, Robert S. "The Journal of Robert S. Bliss." *Utah Historical Quarterly* 4:3 (July 1931): 67–96.

Bolton, Herbert Eugene. *The Padre on Horseback: A Sketch of Eusebio Francisco Kino, S.J., Apostle to the Pimas.* San Francisco: Sonora Press, 1932; reprint, Chicago: Loyola University Press, 1986.

——. *Rim of Christendom: A Biography of Eusebio Francisco Kino, Pacific Coast Pioneer.* New York: Macmillan Company, 1936.

Bowden, Charles. "Death of the Santa Cruz Calls for a Celebration." *Arizona Daily Star,* August 23, 1981.

——. *Killing the Hidden Waters.* Austin: University of Texas Press, 1977.

Brandt, Herbert. *Arizona and Its Bird Life: A Naturalist's Adventures with the Nesting Birds on the Deserts, Grasslands, Foothills, and Mountains of Southeastern Arizona.* Cleveland: Bird Research Foundation, 1951.

Brewster, Jane Wayland. "The San Rafael Cattle Company: A Pennsylvania Enterprise in Arizona." *Arizona and the West* 8:2 (Summer 1966): 138–149.

Brooks, Laura. "Human Demands Put Strain on Flow." *Arizona Daily Star,* October 4, 1994.

——. "River Sustains But Can Isolate Sonora Town." *Arizona Daily Star,* October 2, 1994.

——. "Sonora Leg Nourishes Rural Lives." *Arizona Daily Star,* October 3, 1994.

Browell, Elizabeth R. "The Presido of Tubac." *Voices from the Pimeria Alta.* Nogales, Ariz.: Pimeria Alta Historical Society, 1991.

Browne, J. Ross. *Adventures in the Apache Country: A Tour through Arizona and Sonora, with Notes on the Silver Regions of Nevada.* New York: Harper and Brothers, 1869; reprint, New York: Arno Press, 1973.

Bryan, Alan L. *Paleo-American Prehistory.* Occasional Papers of the Idaho State University Museum No. 16, 1965.

Burkham, D. E. *Depletion of Streamflow by Infiltration in the Main Channels of the Tucson Basin, Southeastern Arizona.* U.S.G.S. Water-Supply Paper 1939-B, 1970.

Burrus, Ernest J. *Kino and Manje: Explorers of Sonora and Arizona; Their Vision of the Future, A Study of Their Expeditions and Plans, with an Appendix of Thirty Documents.* St. Louis: Jesuit Historical Institute, 1971.

Butzer, Karl. *Early Hydraulic Civilization in Egypt: A Study in Cultural Ecology.* Chicago: University of Chicago Press, 1976.

———. *Environment and Archeology: An Ecological Approach to Prehistory.* Chicago: Aldine-Atherton, 1971.

Campa, Arthur L. *Hispanic Culture in the Southwest.* Norman: University of Oklahoma Press, 1979.

Castetter, Edward F., and Willis H. Bell. *Pima and Papago Indian Agriculture.* Albuquerque: University of New Mexico Press, 1942.

Chamberlain, Samuel E. *My Confession.* New York: Harper and Brothers, 1956.

Charter and Ordinances of the City of Tucson. Ben C. Hill, ed. Published by the City of Tucson, 1926.

Clarke, A. B. *Travels in Mexico and California.* Boston: Wright and Hasty Printers, 1852.

Coate, Charles. "The Biggest Water Fight in American History." *Journal of the Southwest* 37:1 (Spring 1995): 79–101.

Coggeshall, Marc Charles. "Hydrologic Assessment and Computer Model Application in the Upper Santa Cruz River Basin, Santa Cruz County, Arizona." M.S. thesis, University of Arizona, 1990.

Cooke, Philip St. George. *The Conquest of New Mexico and California in 1846–1848.* New York: G. P. Putnam's Sons, 1878; reprint Chicago: Rio Grande Press, 1964.

Cooke, Ronald U., and Richard W. Reeves. *Arroyos and Environmental Change in the American South-West.* Oxford: Clarendon Press, 1976.

Couts, Cave Johnson. *Hepah, California! The Journal of Cave Johnson Couts, from Monterey, Nuevo Leon, Mexico to Los Angeles, California during the Years 1848–1849.* Henry F. Dobyns, ed. Tucson: Arizona Pioneers' Historical Society, 1961.

Cox, Cornelius C. "From Texas to California in 1849." Mabelle Eppard Martin, ed. *Southwestern Historical Quarterly* 29:2 (October 1925): 128–146.

Crown, Patricia L. and W. James Judge, eds. *Chaco and Hohokam: Prehistoric Regional Systems in the American Southwest.* Santa Fe: School of American Research Press, 1991.

Cumming, Douglas. "A Rancher Looks at His Environment." *Voices from the Pimeria Alta.* Nogales, Ariz.: Pimeria Alta Historical Society, 1991.

Cushman, R. L. "Lower Santa Cruz Area, Pima and Pinal Counties." *Groundwater in the Gila Basin and Adjacent Areas.* Leonard C. Halpenny, ed. U.S.G.S., Open-File Report, Tucson, Arizona, October 1952.

Czaplicki, Jon S., and James D. Mayberry. *An Archaeological Assessment of the Middle Santa Cruz River Basin, Rillito to Green Valley, Arizona, For the Proposed Tucson Aqueduct Phase B, Central Arizona Project.* University of Arizona, Arizona State Museum, Archaeological Series No. 164, 1983.

Davidson, E. S. *Geohydrology and Water Resources of the Tucson Basin, Arizona.* U.S.G.S. Water Supply Paper 1939–E, 1973.

Davis, Judith C. "The Hydrology and Plant Community Relations of Canelo Hills Cienega, An Emergent Wetland in Southeastern Arizona." M.S. thesis, University of Arizona, 1993.

Davis, Mike. "Las Vegas versus Nature." *Reopening the American West.* Hal K. Rothman, ed. Tucson: University of Arizona Press, 1998.

Dean, Jeffrey S., and William J. Robinson. *Dendroclimatic Variability in the American Southwest, A.D. 680–1970.* University of Arizona, Laboratory for Tree-Ring Research, 1978.

de Castillo, Jose. Manuscript Collection, Arizona Historical Society, MS 140.

DiPeso, Charles. *The Upper Pimas of San Cayetano del Tumacacori: An Archaeohistorical Reconstruction of the Ootam of Pimeria Alta.* Dragoon, Ariz.: Amerind Foundation, 1956.

Dobyns, Henry F. *From Fire to Flood: Historic Human Destruction of Sonoran Desert Riverine Oases.* Socorro, New Mexico: Ballena Press, 1981.

——. "Indian Extinction in the Middle Santa Cruz River Valley, Arizona." *New Mexico Historical Review* 38:1 (January 1963): 163–181.

——. "Prologue." Cave Johnson Couts, *Hepah California! The Journal of Cave Johnson Couts, from Monterey, Nuevo Leon, Mexico to Los Angeles, California During the Years 1848–1849.* Henry F. Dobyns, ed. Tucson: Arizona Pioneers' Historical Society, 1961.

——. *Spanish Colonial Tucson: A Demographic History.* Tucson: University of Arizona Press, 1976.

——. *Tubac through Four Centuries: An Historical Resume and Analysis.* The Arizona State Parks Board, 15 March 1959. University of Arizona Library, microfilm No. 1045.

Doelle, William H. "Human Use of the Santa Cruz River in Prehistory." Paper presented at the American Society for Environmental History Conference, April 14–18, 1999, Tucson, Arizona.

Doelle, William H., Frederick W. Huntington, and Henry D. Wallace. "Rincon Phase Reorganization in the Tucson Basin." *The Hohokam Village: Site, Structure, and Organization.* David Doyel, ed. Glenwood Springs, Colo.: Southwestern and Rocky Mountain Division of the American Association for the Advancement of Science, 1987.

Doelle, William H., and Henry D. Wallace. "The Changing Role of the Tucson Basin in the Hohokam Regional System." *Exploring the Hohokam.* George J. Gumerman, ed. Albuquerque: University of New Mexico Press, 1991.

Downing, Theodore E., and McGuire Gibson, eds. *Irrigation's Impact on Society.* University of Arizona, Anthropological Papers No. 25, 1974.

Downum, Christian E. "Between Desert and River: Hohokam Settlement and Land Use in the Los Robles Community." University of Arizona, Anthropological Papers No. 57, 1993.

Drake, Charles R. Manuscript Collection, Arizona Historical Society, MS 228.

Drew, Linda G. *Tree-Ring Chronologies of Western America.* Vol. 2: *Arizona, New Mexico and Texas.* University of Arizona, Laboratory for Tree-Ring Research, 1972.

Duffen, William A., ed. "Overland via 'Jackass Mail' in 1858: The Diary of Phocion R. Way." *Arizona and the West* 2:1–4 (Spring–Winter 1960).

Durivage, John E. *Daily Picayune* (New Orleans), August 7, 1849.

Eccleston, Robert. *Overland to California on the Southwestern Trail, 1849: Diary of Robert Eccleston.* Edward H. Howes, ed. Berkeley and Los Angeles: University of California Press, 1950.

Emory, William H. *Report on the United States and Mexican Boundary Survey.* Executive Document No. 135, 34th Congress, 1st Session, 1856; reprint, Austin: Texas State Historical Association, 1987.

Euler, Robert C., George J. Gumerman, Thor N. V. Karlstrom, Jeffrey S. Dean, and Richard H. Hevly. "The Colorado Plateaus: Cultural Dynamics and Paleoenvironment." *Science* 205:4411 (September 14, 1979): 1089–1101.

Evans, George W. B. *Mexican Gold Trail: The Journal of a Forty-Niner.* Glenn S. Dunke, ed. San Marino, Calif.: Huntington Library, 1945.

Ezell, Paul H. "Is There a Hohokam Continuum?" *American Antiquity* 29 (July 1963): 61–66.

Fenneman, Nevin M. *Physiography of Western United States.* New York: McGraw-Hill, 1931.

Fergusson, Major David. "Cultivated Fields." 1862 map, Arizona Historical Society, Case 3, Dr. 3, No. 456.

——. "Letter of the Secretary of War in Answer to a Resolution of the Senate, a Copy of the Report of Major D. Fergusson on the Country, Its Resources, and the Route between Tucson and Lobos Bay." 37th Congress, Special Session, Senate Executive Document No. 1, March 14, 1863.

Field, John, Keith Katzer, Jim Lombard, and Jeanette Schuster. "A Geomorphic Survey of the Picacho and Northern Tucson Basins." *The Northern Tucson Basin Survey: Research Directions and Background Studies.* John H. Madsen, Paul R. Fish, and Suzanne K. Fish, eds. University of Arizona, Arizona State Museum, Archaeological Series No. 182, 1993.

Fish, Paul R., and Suzanne K. Fish. "Hohokam Political and Social Organization."

Exploring the Hohokam. George J. Gumerman, ed. Albuquerque: University of New Mexico Press, 1991.

Fish, Suzanne K., Paul R. Fish, Charles H. Mikisicek, and John H. Madsen. "Prehistoric Agave Cultivation in Southern Arizona." *Desert Plants* 7:2. University of Arizona Press, for the Boyce Thompson Southwestern Arboretum, 1985.

Fish, Suzanne K., and Gary Nabhan. "Desert as Context: The Hohokam Environment." *Exploring the Hohokam*. George J. Gumerman, ed. Albuquerque: University of New Mexico Press, 1991.

"Flood Insurance Study, Pima County, Arizona and Incorporated Areas." Vol. 1, revised February 8, 1999. Federal Emergency Management Agency.

Flores, Dan. "Place: An Argument for Bioregional History." *Environmental History Review* 18 (Winter 1994): 1–18.

Fontana, Bernard L. "Calabazas of the Rio Rico." *The Smoke Signal* 24 (Fall 1971): 66–79.

———. *Of Earth and Little Rain: The Papago Indians*. Tucson: University of Arizona Press, 1989.

Foreman, S. W. Maps and Notes, 1871. Townships 14 and 15 South, Range 13 East. Microfiche, Bureau of Land Management Office, Tucson, Arizona.

"A Framework for Guiding the Future of the San Rafael Valley." Tucson: Sonoran Institute.

Frenkel, Stephen. "Old Theories in New Places? Environmental Determinism and Bioregionalism." *Professional Geographer* 46 (August 1994): 289–295.

Froebel, Julius. *Seven Years Travel in Central America, Northern Mexico, and the Far West of the United States*. London: Richard Bentley Publishers, 1859.

Gelt, Joe, Jim Henderson, Kenneth Seasholes, Barbara Tellman, and Gary Woodard. *Water in the Tucson Area: Seeking Sustainability*. University of Arizona, College of Agriculture, Water Resources Research Center, Issue Paper No. 20, 1999.

Goetzmann, William H. "Introduction." William H. Emory, "Report on the United States and Mexican Boundary Survey." Reprint, Austin: Texas State Historical Association, 1987.

Golder, Frank Alfred. *The March of the Mormon Battalion from Council Bluffs to California, Taken from the Journal of Henry Standage*. New York: Century Company, 1928.

Goodman, John G., III, ed. *Personal Recollections of Harvey Wood*. Pasadena, Calif.: 1955.

Graham, James Duncan. *Report of Lt. Col. Graham on the Subject of the Boundary Line between the United States and Mexico*. Senate Executive Document 121, 32nd Congress, 1st Session, 1853.

Greenleaf, J. Cameron. *Excavations at Punta de Agua in the Santa Cruz River Basin, Southeastern Arizona*. University of Arizona, Anthropological Papers No. 26, 1975.

Gregonis, Linda M., and Lisa W. Huckell. *The Tucson Urban Study*. Cultural Resource Management Section, Archaeological Series No. 138, Arizona State Museum, 1980.

Gregonis, Linda M., and Karl J. Reinhard. *Hohokam Indians of the Tucson Basin*. Tucson: University of Arizona Press, 1979.

Gregory, David A. "The Morphology of Platform Mounds and the Structure of Classic Period Hohokam Sites." *The Hohokam Village: Site, Structure, and Organization*. David Doyel, ed. Glenwood Springs, Colo.: Southwestern and Rocky Mountain Division of the American Association for the Advancement of Science, 1987.

Gustafson, A. M, ed. *John Spring's Arizona*. Tucson: University of Arizona Press, 1966.

Hadley, Diana, and Thomas E. Sheridan. *Land Use History of the San Rafael Valley, Arizona (1540–1960)*. Fort Collins, Colo.: U.S. Forest Service, General Technical Report RM-GTR-269, 1995.

Halpenny, Leonard C. *Groundwater in the Gila Basin and Adjacent Areas—A Summary*. U.S.G.S., "Open-File Report—Not reviewed for conformance with editorial standards and stratigraphic nomenclature of the Geologic Survey," Tucson, Arizona, October 1952.

Hammond, George. "Pimeria Alta After Kino's Time." *New Mexico Historical Review* 4:3 (1929): 227–235.

Hastings, James R. "People of Reason and Others: The Colonization of Sonora to 1776." *Arizona and the West* 3:4 (Winter 1961): 321–340.

Hastings, James Rodney, and Raymond M. Turner. *The Changing Mile: An Ecological Study of Vegetation Change With Time in the Lower Mile of an Arid and Semiarid Region*. Tucson: University of Arizona Press, 1965.

Haury, Emil W. *The Hohokam: Desert Farmers and Craftsmen; Excavations at Snaketown, 1964–1965*. Tucson: University of Arizona Press, 1976.

Helmick, Walter Robert. "The Santa Cruz River Terraces Near Tubac, Santa Cruz County, Arizona." M.S. thesis, University of Arizona, 1986.

Hewitt, James M., and David Stephen. "Archaeological Investigations in the Tortolita Mountains Region, Southern Arizona." Pima Community College, Anthropological Series Archaeological Field Report No. 10, 1981.

Hicks, R. B., F. N. Owens, D. R. Gill. J. J. Martin, and C. A. Strasia. "Water Intake by Feedlot Steers." Animal Sciences Research Report, Oklahoma Agricultural Station, 1988.

Hinderlider, M. C. "Irrigation of Santa Cruz Valley, Recovery of Underground Water in Arizona by Wells and Pumps." *Engineering Record* Vol. 68, No. 8 and No. 9, August 23 and August 30, 1913.

Horton, Ken. Interview, May 24, 2000. Notes in the possession of author.

Hughes, Samuel. Manuscript Collection, Arizona Historical Society, MS 366.

Hundley, Norris, Jr. *The Great Thirst: Californians and Water, 1770s-1990s.* Berkeley: University of California Press, 1992.

Ingram, Helen, Nancy K. Laney, and David M. Gillilan. *Divided Waters: Bridging the U.S.-Mexico Border.* Tucson: University of Arizona Press, 1995.

James, George Wharton. *Arizona, The Wonderland.* Boston: Page Company, 1917.

Johnson, P. W. "Upper Santa Cruz Basin, Pima and Santa Cruz Counties." *Groundwater in the Gila Basin and Adjacent Areas.* Leonard C. Halpenny, ed. U.S.G.S., Open-File Report, Tucson, Arizona, October 1952.

Johnson, R. Bruce. "Annual Static Water Level Basic Data Report, Tucson Basin and Avra Valley, Pima County, Arizona, 1994." City of Tucson: Tucson Water Planning and Engineering Division, 1996.

Johnson, Rich. *The Central Arizona Project, 1918–1968.* Tucson: University of Arizona Press, 1977.

Jones, Percy. Interview 1973. University of Arizona Special Collections, Tape 215.

Kappel, Wayne. "Irrigation Development and Population Pressure." Theodore E. Downing and McGuire Gibson, eds. *Irrigations Impact on Society.* University of Arizona, Anthropological Papers No. 25, 1974.

Kelly, Isabel. *The Hodge's Ruin: A Hohokam Community in the Tucson Basin.* University of Arizona, Archaeological Series No. 30, 1978.

Kessell, John L. *Friars, Soldiers and Reformers: Hispanic Arizona and the Sonoran Mission Frontier, 1767–1856.* Tucson: University of Arizona Press, 1976.

———. *Mission of Sorrows: Jesuit Guevavi and the Pimas, 1691–1767.* Tucson: University of Arizona Press, 1970.

———. "The Puzzling Presidio: San Felipe de Guevavi, alias Terrenate." *New Mexico Historical Review* 41:1 (January 1966): 21–46.

Kupel, Douglas E. "Diversity Through Adversity: Tucson Basin Water Control Since 1854." M.A. thesis. University of Arizona, 1986.

Lazaroff, David Wentworth. *Sabino Canyon: The Life of a Southwestern Oasis.* Tucson: University of Arizona Press, 1993.

Logan, Michael F. *Fighting Sprawl and City Hall: Resistance to Urban Growth in the Southwest.* Tucson: University of Arizona Press, 1995.

Long, Jack, interview. Director, Hohokam Irrigation District. Coolidge, Arizona, May 26, 2000. Notes in possession of the author.

MacCameron, Robert, "Environmental Change in Colonial New Mexico." Char Miller and Hal Rothman, eds. *Out of the Woods: Essays in Environmental History.* Pittsburgh: University of Pittsburgh Press, 1997.

MacNeish, Richard S. "Early Man in the New World." *American Scientist* 64 (1976): 316–327.

McCarty, Kieran. *Copper State Bulletin.* Arizona State Genealogical Society XVII:4 (Winter 1982): 91–93.

———. *Desert Documentary: The Spanish Years, 1767–1821.* Tucson: Arizona Historical Society, Historical Monograph No. 4, 1976.

——, ed. *A Frontier Documentary: Sonora and Tucson, 1821–1848*. Tucson: University of Arizona Press, 1997.

——. *A Spanish Frontier in the Enlightened Age: Franciscan Beginnings in Sonora and Arizona*. Washington: Academy of American Franciscan History, 1981.

McEachern, Ron, interview. Director, Central Arizona Irrigation District. Eloy, Arizona. May 26, 2000.

McGuire, Randall H., and Michael B. Schiffer, eds. *Hohokam and Patayan Prehistory of Southwestern Arizona*. New York: Academic Press, 1982.

McKee, Jack Edward, and Harold W. Wolf, eds. "Water Quality Criteria," 2nd Edition. The Resources Agency of California, State Water Quality Control Board, Publication No. 3–A, 1963.

McTaggart, W. Donald. "Bioregionalism and Regional Geography: Place, People, and Networks." *Canadian Geographer* 37 (Winter 1993): 307–319.

Manriquez, Manuel Mascarenas. "A Brief History of the Mascarenas Ranch." *Voices from the Pimeria Alta*. Nogales, Ariz.: Pimeria Alta Historical Society, 1991.

Marshall, Thomas Maitland. "St. Vrain's Expedition to the Gila in 1826." *Southwestern Historical Quarterly* 19:3 (January 1916): 251–260.

Martin, Douglas D. *The Lamp in the Desert: the Story of the University of Arizona*. Tucson: University of Arizona Press, 1960.

Martin, Paul S. "Pleistocene Overkill." *Natural History* 76 (December 1967): 32–38.

Martin, Paul, and Richard Klein, eds. *Quaternary Extinctions: A Prehistoric Revolution*. Tucson: University of Arizona Press, 1984.

Mason, Doug, interview. Director, San Carlos Irrigation and Drainage District. Florence, Arizona, May 26, 2000.

Matson, Daniel S., and Bernard L. Fontana. *Friar Bringas Reports to the King*. Tucson: University of Arizona Press, 1977.

Mattison, Ray H. "Early Spanish and Mexican Settlements in Arizona." *New Mexico Historical Review* 21:4 (1946): 285–327.

Meyer, Michael. *Water in the Hispanic Southwest*. Tucson: University of Arizona Press, 1984.

Miller, Zane L. *The Urbanization of Modern America: A Brief History*. New York: Harcourt Brace Jovanovich, Inc., 1973.

Montes, Leandro. "Effects of the October 1983 Discharge on Channel Configuration in Rillito Creek, Tucson, Arizona." *Journal of the Arizona-Nevada Academy of Science*. Vol. 19, Annual Conference Report, April 1984.

Murbarger, Nell. *Ghosts of the Adobe Walls*. Tucson: Treasure Chest Publications, 1964.

Myrick, David F. *Railroads of Arizona*. Vol. 1: *The Southern Roads*. Berkeley: Howell-North Books, 1975.

Nevins, Allan. *Frémont: The West's Greatest Adventurer*. 2 Vols. New York: Harper and Brothers, 1928.

Nials, Fred L., David A. Gregory, and Donald A. Graybill. "Salt River Streamflow and Hohokam Irrigation Systems." *The 1982–1984 Excavations at Las Colinas: Environment and Subsistence.* 6 vols. Carol Ann Heathington and David A Gregory, eds. University of Arizona, Archaeological Series No. 162, 1989.

Oblasser, Bonaventure. "Papaguería: The Domain of the Papagos." *Arizona Historical Review* 7 (April, 1936): 3–9.

Officer, James E. *Hispanic Arizona, 1536–1856.* Tucson: University of Arizona Press, 1987.

Olberg, C. R., and F. R. Schanck. "Irrigation and Flood Protection, Papago Indian Reservation, Arizona." U.S. 62nd Congress, 3rd Session, Senate Document No. 973, 1913.

Pancoast, Charles Edward. *A Quaker Forty-Niner: The Adventures of Charles Edward Pancoast on the American Frontier.* Anna Paschall Hannum, ed. Philadelphia: University of Pennsylvania Press, 1930.

Parke, John G. "Report on Explorations for that Portion of a Railway Route, Near the Thirty-second Parallel of Latitude, Lying Between Dona Ana, on the Rio Grande, and Pima Villages, on the Gila." House Document 129, U.S. Army Corps of Topographical Engineers, Report Submitted August 22, 1854.

Parker, John T. C. "Channel Change on the Santa Cruz River, Pima County, Arizona, 1936–1986." U.S.G.S., Open-File Report 93–41, December 1993.

Parker, Robert Wade. "Gravity Analysis of the Subsurface Structure of the Upper Santa Cruz Valley, Santa Cruz County, Arizona." M.S. thesis, University of Arizona, 1978.

Parsons, James J. "On 'Bioregionalism' and 'Watershed Consciousness.' " *The Professional Geographer* 37 (February 1985): 1–6.

Patagonia-Sonoita Creek Preserve. Informational Brochure, May 2000. Available from the Patagonia-Sonoita Creek Preserve, P.O. Box 815, Patagonia, Arizona 85624.

Pederson, Gilbert J. "A Yankee in Arizona: The Misfortunes of William S. Grant, 1860–1861." *Journal of Arizona History* 16:4 (Summer 1975): 127–144.

Pence, Angelica. "Arid Ditch in Tucson Belies Past." *Arizona Daily Star,* October 5, 1994.

Pence, Angelica, and Laura Brooks. "Santa Cruz: On the River." *Arizona Daily Star,* October 2, 1994.

Peterson, Thomas H. Jr. "Fort Lowell, A.T., Army Post During the Apache Campaigns." *The Smoke Signal,* "Published occasionally by The Tucson Corral of the Westerners." Number 8 (Fall 1963): 1–19.

Pfefferkorn, Ignaz. *Sonora: Descriptions of the Province.* Theodore E. Treutlein, trans. Tucson: University of Arizona Press, 1989. Originally published in *The Coronado Cuarto Centennial Publications, 1540–1940.* Vol. 12. George P. Hammond, ed. Albuquerque: University of New Mexico Press, for the Coronado Historical Fund, 1949.

Pisani, Donald J. *Water, Land, and Law in the West: The Limits of Public Policy, 1850–1920.* Lawrence: University of Kansas Press, 1996.

Polzer, Charles W. *Kino Guide II: His Missions—His Monuments.* University of Arizona, Arizona State Museum, Southwestern Mission Research Center, 1982.

Ponce, Victor Miguel, Zbig Osmolski, and Dave Smutzer. "Large Basin Deterministic Hydrology: A Case Study." *Journal of Hydraulic Engineering.* 111:9 (September 1985): 1227–1245.

Poston, Charles D. *Building a State in Apache Land.* John M. Meyers, ed. Tempe, Ariz.: Aztec Press, Inc. 1963.

Potash, Robert A. "Notes and Documents." *New Mexico Historical Review* 24:4 (October 1949): 332–335.

Powell, H.M.T. *The Santa Fe Trail to California: The Journal and Drawings of H.M.T. Powell, 1849–1852.* Douglas S. Watson, ed. San Francisco: Book Club of California, 1931.

"A Profile of Arizona's San Rafael Valley." The Sonoran Institute, 7290 E. Broadway, Tucson, Arizona, 85710.

Random House Dictionary of the English Language. 2nd edition, Unabridged. New York: Random House, 1987.

Reisner, Marc. *Cadillac Desert: The American West and Its Disappearing Water.* New York: Viking Penguin, Inc., 1986.

de Ribas, Andres Perez. *My Life Among the Savage Nations of New Spain.* Thomas Robertson, trans. Los Angeles: Ward Ritchie Press, 1968.

Robinson, William J. "Mission Guevavi: Excavations in the Convento." *The Kiva* 42:2 (1976): 135–175.

Rodriguez, Alberto. "The Otro Lado—Nogales, Sonora." *Voices from the Pimeria Alta.* Nogales, Ariz.: Pimeria Alta Historical Society, 1991.

Roeske, R. H., J. M. Garrett, and J. H. Eychaner. "Floods of October 1983 in Southeastern Arizona." U.S.G.S. Water Resources Investigations Report 85-4225-C, 1989.

Rothrock, J. T. "Preliminary Botanical Report, With Remarks Upon the General Topography of the Region Traversed in New Mexico and Arizona; Its Climatology, Forage-Plants, Timber, Irrigation, Sanitary Conditions, etc." United States Engineer Office, Geographical Explorations and Surveys West of the One Hundredth Meridian, 1875.

Russell, Frank. "The Pima Indians." *Twenty-sixth Annual Report of the Bureau of American Ethnology, 1904–1905.* U.S. Government Printing Office, 1908.

Saarinen, Thomas F., Victor R. Baker, Robert Durrenberger, and Thomas Maddock, Jr. *The Tucson, Arizona, Flood of October 1983.* Washington: National Academy Press, 1984.

Safford, A.P.K., and Samuel Hughes. "The Story of Marianna Diaz." In the Safford Collection, Arizona Historical Society. MS 704.

Schwalen, Harold Christy. Papers, 1919–1965. University of Arizona Library Special Collections, AZ 563.

Schwalen H. C., and R. J. Shaw. "Ground Water Supplies of Santa Cruz Valley of Southern Arizona Between Rillito Station and the International Boundary." University of Arizona, Agricultural Experiment Station Bulletin No. 288, 1957.

Seibold, Frank M. "Patagonia, Then and Now." *Voices from the Pimeria Alta.* Nogales, Ariz.: Pimeria Alta Historical Society, 1991.

Sheridan, Thomas E. *Arizona, A History.* Tucson: University of Arizona Press, 1995.

———. *Los Tucsonenses: The Mexican Community in Tucson, 1854–1941.* Tucson: University of Arizona Press, 1986

Smith, Fay Jackson, John L. Kessell, and Francis J. Fox. *Father Kino in Arizona.* Phoenix: Arizona Historical Foundation, 1966.

Smith, G.E.P. "Groundwater Supply and Irrigation in the Rillito Valley." University of Arizona, Agricultural Experiment Station, Bulletin No. 64, 1910.

———. Papers. University of Arizona Library Special Collections, MS 280.

Sonnichsen, C. L. *Colonel Greene and the Copper Skyrocket* Tucson: University of Arizona Press, 1974.

———. *Tucson: The Life and Times of an American City.* Norman: University of Oklahoma Press, 1982.

Spalding, Volney M. "Distribution and Movements of Desert Plants." Washington, D.C.: Carnegie Institute of Washington Publication No. 113, 1909.

Stokes, William Lee, and Sheldon Judson. *Introduction to Geology, Physical and Historical.* Englewood Cliffs, New Jersey: Prentice-Hall, 1968.

de la Teja, Jesus F. *San Antonio de Bexar: A Community on New Spain's Northern Frontier.* Albuquerque: University of New Mexico Press, 1995.

Tellman, Barbara, Richard Yarde, and Mary G. Wallace. "Arizona's Changing Rivers: How People Have Affected the Rivers." University of Arizona, College of Agriculture, Water Resources Research Center, 1997.

Third Management Plan, 2000–2001. Tucson Active Management Area. Arizona Department of Water Resources, December 1999.

Thornber, J. J. "Plant Acclimatization in Southern Arizona." *The Plant World* 14 (January 1911): 1–9.

Tombstone Epitaph.

"The Truth About Water in Tucson." Tucson Chamber of Commerce, 1953. University of Arizona Library Special Collections.

Tucson Citizen.

Tucson Magazine.

U.S. Census Reports.

U.S. Surgeon General's Office. "A Report on Barracks and Hospitals, with Descriptions of Military Posts." Circular No. 4, War Department, Surgeon General's Office, 1870.

Wagoner, Jay J. *Early Arizona: Prehistory to Civil War.* Tucson: University of Arizona Press, 1975.

Warner, Solomon. Manuscript Collection, Arizona Historical Society, MS 844.

Wayland, J. A. "Experiment on the Santa Cruz: Collin Cameron's San Rafael Cattle Company, 1882–1893." M.A. thesis, University of Arizona, 1964.

Webb, Robert H., and Julio L. Betancourt. "Climatic Variability and Flood Frequency of the Santa Cruz River, Pima County, Arizona." U.S.G.S. Water-Supply Paper No. 2379, 1992.

Webb, Walter Prescott. *The Great Plains*. Lincoln: University of Nebraska Press, 1931.

Weber, David J. *The Taos Trappers: The Fur Trade in the Far Southwest, 1540–1846*. Norman: University of Oklahoma Press, 1970.

Weekly Arizonan.

Welsh, Frank. *How to Create a Water Crisis*. Boulder, Colo.: Johnson Publishing Company, 1985.

Wheeler, C. C. "History and Facts Concerning Warner and Silver Lakes and the Santa Cruz River." Arizona Historical Society, MS 853.

Whelan, Jamie. "From Archives and Trash Pits: Sources for Analyzing Food Consumption Patterns in Historic Sites Archaeology." Paper presented at the 39th Annual Conference of the Western Social Sciences Association, Albuquerque, New Mexico, April 23–26, 1997.

White, Lynn. "The Historical Roots of our Ecological Crisis." *Science* 122 (1967): 1203–1207; reprinted in Paul Shepard and Daniel McKinley, eds. *The Subversive Science: Essays Toward An Ecology of Man*. Boston: Houghton Mifflin, 1969.

White, Richard. *"It's Your Misfortune and None of My Own:" A New History of the American West*. Norman: University of Oklahoma Press, 1991.

Whitworth, Robert W. "From the Mississippi to the Pacific: An Englishman in the Mormon Battalion." David B. Gracy II and Helen J. H. Rugeley, eds. *Arizona and the West* 7:2 (Summer 1965): 127–160.

Wilcox, David R., Thomas R. McGuire, and Charles Sternberg. "Snaketown Revisited: A Partial Cultural Resource Survey, Analysis of Site Structure, and an Ethnohistoric Study of the Proposed Hohokam-Pima National Monument." University of Arizona, Arizona State Museum Anthropological Papers No. 23, 1980

Willey, Richard R. "La Canoa: A Spanish Land Grant Lost and Found." *The Smoke Signal, "*Published by the Tucson Corral of the Westerners." 38 (Fall 1979): 153–172.

Willson, Roscoe G. *Pioneer Cattlemen of Arizona*. Phoenix: McGrew Commercial Printery, for the Valley National Bank, 1951.

Wilson, E. D., J. B. Cunningham, and G. M. Butler. "Arizona Lode—Gold Mines and Gold Mining." *Arizona Bureau of Mines Bulletin*. No. 137, 1967.

Wilson, John P. *Islands in the Desert: A History of the Upland Areas of Southeast Arizona*. Albuquerque: University of New Mexico Press, 1995.

Wissler, Clark. *Indians of the United States.* New York: Doubleday and Company, 1966.

Wittfogel, Karl A. "The Hydraulic Approach to Pre-Spanish Mesoamerica." *The Prehistory of Tehuacan Valley.* Vol. 4: *Chronology and Irrigation.* Frederick Johnson, ed. Austin: University of Texas Press, 1972.

Wood, Harvey. *Personal Recollections of Harvey Wood.* Angels Camp, Calaveras County, Calif.: Mountain Echo Job Printing Office, 1878; reprint Pasadena, California, 1955.

Worcester, Donald Emmet. "The Apache Indians of New Mexico in the Seventeenth Century." M.A. thesis, University of California, Berkeley, 1940.

Worster, Donald. *Rivers of Empire: Water, Aridity, and the Growth of the American West.* New York: Oxford University Press, 1985.

Young, R. A. "The Arizona Water Controversy: An Economist's View." Arizona Academy of Science, May 1968.

Young, Robert A., and William E. Martin. "The Economics of Arizona's Water Problem." *Arizona Review* 16:3 (March 1967): 9–18.

Illustration Credits

Index

commerce and trade: with Forty-
Niners, 86, 87–89, 91; in Mexican
era, 68, 75–76, 78; of military colo-
nies, 94, 96; with Mormon Bat-
talion, 80, 82–83; in Spanish era, 42,
54, 55–56; in Tucson, 114; Solomon
Warner and, 122
cones of depression, 188–89, 214
Congress Street (Tucson), 126, 147,
178, 192, 222
conifer ecosystems, 22–24
Consolidated Aircraft Company, 194–
95
Continental, Ariz., 192, 212, 221–22,
230, 256
Continental Rubber Company, 162
Cooke, Col. Philip St. George, 79, 80,
88; and Cooke's Trail, 108
Cooke and Reaves, 140
Coolidge, Ariz., 230, 240
Coronado, Francisco Vásquez de, 42
Corps of Engineers, 190
Cortaro, 214, 220, 221–22
cottonwood ecosystems, 7, 8, 23, 83,
96, 129, 165, 258; destroyed in
1930s, 182; during Hohokam era,
33; as marker of shallow aquifer, 207,
214; observed during 1850s, 103,
105–6, 109; restored by effluent
flows, 209, 238, 259
Couts, Lt. Cave Johnson, 82–85
creosote and sagebrush ecosystems, 8,
249
Culver, J. P., 155–56, 178, 222
Cumming, Douglas, 182–83, 185,
209–10
Cushman, R. L., 187
Czaplicki, Jon S., 44

Dalton, C. A., 144
dams, 94–95, 114, 115, 116; Hoover
Dam, 229, 231; Parker Dam on Col-

orado River, 231; proposed, near
Nogales, 151; proposed, in Sabino
Canyon, 150, 173; and Salt River
Project, 164; on Sonoita Creek
(Lake Patagonia), 209; at Warner's
Lake, 143
Davidson, E. S., 213–14
Davis, Mike, 159, 214, 226
Davis, W. C., 144
Davis-Monthan Airfield, 180, 194
DeConcini, Evo, 200–201
dendrochronology, 32
Depression, Great, 181–83, 188, 205;
and Dust Bowl, 184; and New Deal,
186, 189–90, 194–95, 199, 205
Díaz, Father Rafael, 72–73
dike (1913), 166, 170–71
Dobyns, Henry, 52–53, 60
Doelle, William, 39–40, 61, 254
Doyel, David, 38
Driscoll (hotel and ranch owner), 128,
129, 145
drought, 31, 33, 176; in 1760s, 50; in
1770s, 52; in 1790s, 55; in 1832, 69;
in 1870s, 129; in 1890s, 140, 152,
161, 178; in 1920s, 179, 218; in
1930s, 184; in 1950s, 184–85, 201
Durivage, John E., 86–89
Dusenbery, Katie, 216
Dust Bowl, 184

Early Agricultural Period, 29, 40
Early Ceramic Period, 29
El Grande fissure, 188
Elías, Juan, 69–70, 74–75
El Niño, 20, 75, 98, 130, 140; and 1977
flood, 220; and 1983 flood, 221
El Ojito, 119, 128
Eloy, Ariz., 188, 230, 239–40, 250
El Paso, 102
El Pueblito, 52, 55, 66–67, 71, 73–74,
77, 84, 103

Guevavi Ranch, 242
Gulf of California, 117

Halpenny, Leonard C., 195–97
Harshaw Canyon, 139
Haury, Emil, 30, 36
Hayden, Sen. Carl, 186, 231–32
headcuts, 170, 179; in Hohokam era, 34–35; Sam Hughes and arroyo formation of, 145–49; on Valencia Road in Tucson, 143, 151–52
Heady and Ashburn Ranch, 160, 184–86
Hell's Hollow, 141
Helmick, David, 19, 21
Heritage Fund, 241
Hermosillo, Mexico, 96
Herrán, Francisco, 66–67
"Highline Canal," 229, 231
Hohokam, 9, 28–42, 45, 70; alteration of river channel by, 34–35, 41, 59; archaic perception by, 34, 43–44, 207; "Classic" period of, 31, 38, 39; disappearance of, 30, 40; origin of, 29, 30; pithouses of, 37; platform mounds of, 37, 38–39; population estimates for, 30, 36–40, 61
Homer, Porter, 198
Hooker, Henry, 124
Hoover, Herbert: as secretary of commerce, 229
Huachuca Mountains, 82, 139
Hughes, Sam, 115–16, 124, 144, 145–49, 178
Hundley, Norris, 159
Hunt, Gov. George W. P., 229, 231
hydraulic societies, 4, 164–65, 175

Imperial Valley, 231
international boundary: Mexico and, 5, 59, 82, 93, 112, 259; and Water Commission, 257

Interstate 10, 102, 240, 250, 254
Interstate 19, 221, 254, 256–57
irrigation districts, 210–11; Avra-Marana, 187; Casa Grande-Florence-Sacaton, 187; Central Arizona (Eloy), 239–40, 250; Cortaro–Cañada del Oro, 188; Eloy, 187; Flowing Wells, 169; Hohokam (Coolidge), 240
Irrigation Nonexpansion Areas, 225

Janos, 107
Jeffords, T. J., 128
Jesuit explusion, 50, 51
Johnson, Pres. Lyndon B., and Great Society, 232
Johnson, P. W., 196
Joshua trees, 22

Kearney, Gen. Stephen, 78–79, 110
Kennedy, Doug, 216
Kessell, John, 43, 45–46, 50
Ki-He-Kah Ranch, 236
Kino, Father Esebio Francisco, 43–44, 46–47, 72, 76
Kino Springs, Ariz., 258, 260
Kitchen, Pete, 124

Lake Patagonia, 209, 257
landfills, 18–19, 192, 201
land grants, 59, 62; San Ignacio de la Canoa, 62, 70; San Rafael de la Zanja, 159
Leatherwood, Robert, 151
Lee's Ferry, Ariz., 230
León, Manuel de, 56–57
Lindbergh, Charles, 180
Lochiel, Ariz., 70, 236, 259
Los Angeles, 231
lost balance assertion and theory, 156, 181, 191–93, 197

by Indians, 50, 52, 70–71; during
1870s, 130; during 1890s, 140; and
floods, 25, 148–49; during Spanish
era, 45, 53
"overplus lands," 63, 138, 160
Owen brothers, 137–38. *See also*
Walker family

Paleo-Indians, 15, 21–26, 28, 65; and
Pleistocene Overkill theory, 25
Pantano Wash, 28, 173, 226
Papago Indians. *See* Tohono O'odham
"Papago wine," 35, 48, 74
Parke, Lt. John G., 108, 126
Parker, James, 161
Patagonia, Ariz., 206, 221, 230
Patagonia Mountains, 17, 122, 139
Patagonia–Sonoita Creek Preserve,
207, 230
Pattie, James Ohio, 65–66
Pearl Harbor, 193–94
Peck Canyon, 138, 220
peninsulares, 66
perception of Santa Cruz River,
archaic (premodern), 10, 27; by
Anglos, 70, 105, 115, 142, 156, 177,
193; by the Hohokam, 34, 41; by the
Spanish, 43, 46, 58–59; persistance
during postmodern era, 206, 207,
244–45
perception of Santa Cruz River, modern, 10, 33–34, 93, 108, 113, 122–
23, 180; in 1870s, 124, 126, 131; in
1880s, 135–36, 138, 142; in 1890s,
153, 156–57; in 1940s, 192–93; during postmodern era, 206, 226, 229,
233–35, 245–46, 251
perception of Santa Cruz River, postmodern, 9, 10–11, 206, 221; CAP
canal and, 229, 251; disassociation,
180, 248; in 1950s, 201; lack of political consensus, 219, 234, 246–47

Pérez, Ignacio, 63
Pete Kitchen Ranch, 124
Pfefferkorn, Father Ignacio, 48
Phoenix, 3, 66, 139, 240, 249–50; CAP
canal and contracts, 229, 233, 242;
population boom, 240
Phoenix Basin, 18; as center of Hoho-
kam culture, 31–32, 38–39, 42
Phy, Joseph, 120, 151
Picacho Peak, 23, 65, 177, 188, 250–
51; references to, in 1850s, 102–3,
108, 111, 121
Pima County Board of Supervisors,
215–16, 226, 243
Pima Farms, 169
Pima Indians, 38, 111; during 1840s,
74, 84–85; during Spanish era, 43,
52, 65, 67; and Pima Revolt, 46
Pinal Apaches, 59
pinole, 35–36, 48, 80, 89
pinyon pine and juniper ecosystem,
21–24
Pioneer Mill, 128
Pisani, Donald, 145
playa basins, 18
Pleistocene mammals, 22, 24–26;
Pleistocene Overkill theory on, 25,
26, 28
pollution, 53, 238, 242; aircraft industry and, 195; as border issue, 237
Post, Edwin R., and Post Project, 169
postmodern perception. *See* perception
of the Santa Cruz River,
postmodern
Poston, Charles, 118–19, 121
precipitation cycles, 31, 156–57; during Hohokam era, 31, 40; during
Spanish era, 50, 75; in 1880s, 3, 143;
Mormons in Tubac and, 104–5; as
"normal," 124–25, 129
Preston, Daniel, 245
prickly pear cactus, 33

Wilson, James, 123
windmills, 129, 139, 150, 184
wine, 58
Winterhaven, 225
Wood, Harvey, 86–87
Works Progress Administration. *See*
 Depression, Great; New Deal
World War I, 162, 169, 177

World War II, 181, 193–95, 205; Pearl
 Harbor, 193–94
Worster, Donald, 4, 164

Young, Robert A., 216
yucca cactus, 22

Zúñiga, José de, 56–58, 73

About the Author

MICHAEL LOGAN, a native of southern Arizona, received his Ph.D. in U.S. History from the University of Arizona in 1994. He is Professor of History at Oklahoma State University, teaching courses in environmental, urban, U.S. West, and modern U.S. history. He is the author of *Fighting Sprawl and City Hall* (University of Arizona Press, 1995) and *Desert Cities: The Environmental History of Phoenix and Tucson* (University of Pittsburgh Press, 2006). He is currently working on an urban–environmental history of Tucson and Phoenix.